COMPOSTING

YARD AND MUNICIPAL SOLID WASTE

U.S. ENVIRONMENTAL PROTECTION AGENCY
Office of Solid Waste and Emergency Response

CRC Press is an imprint of the
Taylor & Francis Group, an informa business

Reprinted 2010 by CRC Press

CRC Press
6000 Broken Sound Parkway, NW
Suite 300, Boca Raton, FL 33487
270 Madison Avenue
New York, NY 10016
2 Park Square, Milton Park
Abingdon, Oxon OX14 4RN, UK

HOW TO ORDER THIS BOOK
BY PHONE: 800-233-9936 or 717-291-5609, 8AM–5PM Eastern Time
BY FAX: 717-295-4538
BY MAIL: Order Department
Technomic Publishing Company, Inc.
851 New Holland Avenue, Box 3535
Lancaster, PA 17604, U.S.A.
BY CREDIT CARD: American Express, VISA, MasterCard

PERMISSION TO PHOTOCOPY–POLICY STATEMENT
Authorization to photocopy items for internal or personal use, or the internal or personal use of specific clients, is granted by Technomic Publishing Co., Inc. provided that the base fee of US $3.00 per copy, plus US $ .25 per page is paid directly to Copyright Clearance Center, 222 Rosewood Drive, Danvers, MA 01923, USA. For those organizations that have been granted a photocopy license by CCC, a separate system of payment has been arranged. The fee code for users of the Transactional Reporting Service is 1-56676/95 $5.00 + $ .25.

Chapter Overview

Chapter 1 – Planning
Describes the importance of planning and discusses some of the preliminary issues that decision-makers should examine before embarking on any type of composting program including waste characterization, operational plans, facility ownership and management, community involvement, vendors, and pilot programs.

Chapter 2 – Basic Composting Principles
Provides a brief scientific overview of the composting process. Discusses the physical, chemical, and biological factors that influence composting including the type and number of microorganisms present, oxygen level, moisture content, temperature, nutrient levels, acidity/alkalinity, and particle size of the composting material.

Chapter 3 – Collection Methods
Describes options for collecting yard trimmings and municipal solid waste (MSW) along with the advantages and disadvantages associated with each option. Highlights the critical role that source separation plays when composting MSW.

Chapter 4 – Processing Methods, Technologies, and Odor Control
Discusses the three stages of composting (preprocessing, processing, and postprocessing). Introduces the types of equipment associated with each stage, which are examined in detail in Appendix B. Describes the methods currently used to compost yard trimmings and MSW in the United States, and provides a detailed discussion of odor control.

Chapter 5 – Facility Design and Siting
Describes factors to consider when siting and designing a composting facility including location, site topography, and land requirements. Also discusses design considerations for preprocessing, processing, and postprocessing areas; buffer zones; access and onsite roads; and site facilities and security.

Chapter 6 – The Composting Process: Environmental, Health, and Safety Concerns
Focuses on how to prevent or minimize the potential environmental impacts associated with composting including the potential for water pollution, air pollution, odors, vectors, fires, noise, and litter. Discusses the safety and health risks including bioaerosols to workers at composting facilities and ways to minimize these risks.

Chapter 7 – State Legislation and Incentives
Presents an overview of state legislation activity throughout the country. Also discusses state incentives to stimulate yard trimmings and MSW composting.

Chapter 8 – Potential End Users
Describes the potential end users of compost derived from yard trimmings and MSW (agriculture, landscaping, nurseries, silviculture, public agencies, and residents). Discusses how compost is currently utilized by these end users as well as the potential for expanded use.

Chapter 9 – Product Quality and Marketing
Emphasizes the importance of securing markets for the finished compost product. Provides a detailed discussion of quality and safety concerns that could affect the marketability of compost. Also discusses key factors associated with marketing including pricing, distribution, education and public relations, and program assessment.

Chapter 10 – Community Involvement
Discusses the importance of developing strong local support for a composting operation. Also discusses ways to involve and educate the community throughout the planning, siting, operation, and marketing phases of a composting program.

Chapter 11 – Economics
Introduces the economic and financial issues that must be examined when planning a composting facility. Discusses capital costs, operation and maintenance costs, and potential benefits associated with starting up and maintaining a facility.

Appendix A – Additional Sources of Information on Composting
Lists publications related to composting as well as EPA contacts.

Appendix B – Composting Equipment
Describes the cost, efficiency, and major advantages and disadvantages of the equipment commonly used at a composting facility.

Appendix C – Glossary of Compost Terms
Defines terms used throughout the guidebook.

Acknowledgments

EPA would like to thank the following individuals for their contributions to the manual:

Jan Beyea	National Audubon Society
Charlie Cannon	Solid Waste Composting Council
Steve Diddy	Washington State Department of Ecology
Dr. Melvin Finstein	Rutgers University
Dr. Charles Henry	University of Washington
Francine Joyal	Florida Department of Environmental Regulation
Jack Macy	Massachusetts Department of Environmental Protection
Randy Monk	Composting Council
Dr. Aga Razvi	University of Wisconsin
Dr. Thomas Richard	Cornell University
Connie Saulter	Northeast Recycling Council
Dr. Wayne Smith	University of Florida
Roberta Wirth	Minnesota Pollution Control Agency

In addition, EPA thanks Wayne Koser of City Management Corporation and Steve Diddy for their contributions to Appendix B: Composting Equipment.

Contents

	Page
Introduction	1
Composting as a Component of Integrated Solid Waste Management	1
What Is Composting?	2
Status of Composting Yard Trimmings and MSW in the United States	3
Chapter One – Planning	11
Goal Setting	11
Waste Characterization	11
Operational Plans	12
Community Involvement	12
Facility Ownership and Management	13
Composting Vendors	14
Pilot Programs	14
Summary	14
Chapter Two – Basic Composting Principles	16
Overview of the Composting Process	16
The Role of Microorganisms	16
Factors Influencing the Composting Process	17
Oxygen	18
Particle Size	18
Nutrient Levels and Balance	19
Moisture	19
Temperature	19

Table of Contents

Chapter Two – Basic Composting Principles *(Continued)*

    Acidity/Alkalinity (pH) 19

  Summary 20

Chapter Three – Collection Methods 21

  Factors in Yard Trimmings Collection 21

    Public Drop-Off Sites for Yard Trimmings 21

    Curbside Collection of Yard Trimmings 22

  Factors in MSW Collection 28

    Source-Separated MSW 28

    Commingled MSW 29

  Summary 30

Chapter Four – Processing Methods, Technologies, and Odor Control 31

  Preprocessing 31

    Sorting 31

    Reducing the Particle Size of the Feedstock 37

    Treating Feedstock Materials to Optimize Composting Conditions 37

    Mixing 39

  Processing 40

    The Composting Stage 40

    The Curing Stage 47

  Odor Control 47

    Process Control 48

    Engineering Controls 48

  Postprocessing 51

  Summary 53

Chapter Five – Facility Siting and Design 56

  Siting 56

    Location 56

    Topography 59

    Land Area Requirements 59

Chapter Five – Facility Siting and Design (Continued)

        Other Factors Affecting Siting Decisions 60

    Design 60

        Preprocessing Area 60

        Processing Area 60

        Postprocessing Area 62

        Buffer Zone 63

        Access and Onsite Roads 63

        Site Facilities and Security 64

    Summary 64

Chapter Six – The Composting Process: Environmental, Health, and Safety Concerns . . . 65

    Environmental Concerns During Composting 65

        Water Quality 65

        Run-On/Ponding 68

        Air Quality 68

        Odor 70

        Noise 70

        Vectors 70

        Fires 70

        Litter 71

    Occupational Health and Safety Concerns During Composting 71

        Bioaerosols 71

        Potentially Toxic Chemicals 72

        Noise Control 73

        Other Safety Concerns 73

    Worker Training 73

    Summary 74

Chapter Seven – State Legislation and Incentives 76

    Composting Legislation Overview 76

        Permit and Siting Requirements 77

Chapter Seven – State Legislation and Incentives *(Continued)*

    Facility Design and Operations Standards

        Product Quality Criteria

        Bans on Landfilling or Combustion

        Recycling Goals

        Requirements for Local Governments to Implement Composting

        Requirements for State Agencies to Compost

        Separation Requirement

    Yard Trimmings and MSW Composting Incentives

        State Encouragement and Local Authority to Implement Programs

        Grants

        Procurement

        Encouragement of Backyard Composting

        Education Programs

    Summary

Chapter Eight – Potential End Users

    The Benefits of Finished Compost

    Agricultural Industry

    Landscaping Industry

    Horticultural Industry

    Silviculture

    Public Agencies

    Residential Sector

    Summary

Chapter Nine – Product Quality and Marketing

    Product Quality

        Yard Trimmings Compost Quality

        MSW Compost Quality

    Product Specifications 1

Table of Contents

Chapter Nine – Product Quality and Marketing *(Continued)*

    Product Testing 102

    Market Assessment 103

        Private vs. Community Marketing 104

        Pricing 104

        Location/Distribution Issues 107

        Education and Public Relations 108

        Updating the Market Assessment 108

    Summary 108

Chapter Ten – Community Involvement 111

    Planning the Composting Project 111

    Community Involvement in Siting Decisions 112

    Public Participation in the Composting Project 113

    Community Education at the Marketing Phase 114

    Summary 114

Chapter Eleven – Economics 115

    Cost/Benefit Analysis 115

    Capital Costs 116

        Site Acquisition 116

        Site Preparation/Land Improvements 116

        Vehicle and Equipment Procurement 116

        Training 117

        Permits 117

    Operating and Maintenance (O&M) Costs 117

        Collection Costs 117

        Labor Costs 118

        Fuel, Parts, and Supplies 118

        Outreach and Marketing Costs 119

        Other Costs 119

Table of Contents

Chapter Eleven – Economics *(Continued)*

    Benefits From Composting 119

        Avoided Costs 119

        Revenues 119

    Summary 120

Appendix A – Additional EPA Sources of Information on Composting 124

Appendix B – Composting Equipment 126

Appendix C – Glossary of Compost Terms 138

Introduction

> Composting is a form of recycling. Like other recycling efforts, the composting of yard trimmings and municipal solid waste can help decrease the amount of solid waste that must be sent to a landfill or combustor, thereby reducing disposal costs. At the same time, composting yields a valuable product that can be used by farmers, landscapers, horticulturists, government agencies, and property owners as a soil amendment or mulch. The compost product improves the condition of soil, reduces erosion, and helps suppress plant diseases.
>
> The purpose of this manual is to aid decision-makers in planning, siting, designing, and operating composting facilities. It also will be useful to managers and operators of existing facilities, as well as to citizens, regulators, consultants, and vendors interested in the composting process. The manual discusses several approaches to composting and outlines the circumstances in which each method should be considered.
>
> As detailed in the manual, a composting operation should be designed according to the needs and resources of the community. For example, a municipal composting effort can entail simply collecting yard trimmings on a seasonal basis and using a simple "windrow and turn" technology to produce the compost, or it can mean siting and designing a large facility that is capable of handling several tons of mixed municipal solid waste a day.
>
> When considering any type of composting effort, however, decision-makers must plan ahead to avoid potential obstacles that could hinder the operation. The most common challenges are siting the facility, ensuring that the facility is properly designed, mitigating and managing odors, controlling bioaerosols, and investing adequate capital to cover unforeseen costs. This manual helps decision-makers understand and prepare for these challenges so that they can develop a successful composting program in their community.

In 1990, Americans generated over 195 million tons of municipal solid waste (MSW). The amount of waste generated annually in this country has more than doubled in the past 30 years (EPA, 1992). While MSW generation rates have increased, however, the capacity to handle these materials has declined in many areas of the country. Many landfills have closed because they are full. Others are choosing to shut down rather than meet stringent new regulations governing their design and operation. In addition, new landfills and combustors are increasingly difficult to site. In conjunction with this growing gap in disposal capacity, tipping fees at solid waste management facilities are rising in many communities, and the trend does not appear to be changing. As communities search for safe and effective ways to manage MSW, composting is becoming a more attractive management option.

In some communities, composting has proven to be more economical than landfilling, combustion, or constructing new landfills or combustors, especially when considering disposal costs avoided through composting and reduced expenditures on soil amendments for municipal parks and lawns. In addition, composting can help communities meet goals to recycle and divert substantial portions of the MSW stream from disposal. Many states are now setting ambitious recycling goals for their jurisdictions. Because composting can potentially handle up to 30 to 60 percent of a community's MSW stream (EPA, 1993), it can play a key role in helping communities meet these goals. Finally, as a type of recycling, composting in many ways represents a more efficient and a safer use of resources than landfilling or combustion.

Composting as a Component of Integrated Solid Waste Management

EPA encourages communities to use a mix of management techniques (an approach called integrated solid waste management) to handle their MSW stream since no

Introduction

single approach can meet the needs of all communities. EPA suggests a hierarchy of management methods for officials to consider when developing a solid waste management plan. Source reduction is the preferred management option. Source reduction can be defined as the design, manufacture, purchase, or use of materials or products (including packages) to reduce their amount and toxicity before they enter the MSW stream. Recycling, including composting, is the next preferred management option. While lower on the hierarchy than source reduction and recycling, combustion (with energy recovery) and landfilling also are options to manage materials that cannot be reduced, reused, recycled, or composted. Combustion reduces the amount of nonrecyclable materials that must be landfilled and offers the benefit of energy recovery. Landfilling is needed to manage certain types of nonreusable, nonrecyclable materials, as well as the residues generated by composting and combustion.

In any case, consideration of a composting program should be part of a community's comprehensive approach to solid waste management. As decision-makers evaluate their options for managing solid waste, many will look to composting as an attractive and viable option for handling a portion of their MSW stream.

What Is Composting?

Biological decomposition is a natural process that began with the first plants on earth and has been going on ever since. As vegetation falls to the ground, it slowly decays, providing minerals and nutrients needed for plants, animals, and microorganisms. Composting is often used synonymously with biological decomposition. As the term is used throughout this guidebook, however, composting refers to the *controlled* decomposition of organic (or carbon-containing) matter by microorganisms (mainly bacteria and fungi) into a stable humus material that is dark brown or black and has an earthy smell. The process is controlled in that it is managed with the aim of accelerating decomposition, optimizing efficiency, and minimizing any potential environmental or nuisance problems that could develop.

Composting programs can be designed to handle yard trimmings (e.g., leaves, grass clippings, brush, and tree prunings) or the compostable portion of a mixed solid waste stream (e.g., yard trimmings, food scraps, scrap paper products, and other decomposable organics). These materials are the feedstock or "food" for the composting process. Composting programs also have been designed for sewage biosolids, agricultural residues and livestock manures, food processing by-products, and forest industry by-products. Because these materials are not considered part of the MSW stream, however, they are not discussed at length in this guidebook. Some facilities compost

Composting Food Scraps

Each year, a Seattle-area chain of grocery stores called Larry's Markets composts almost 500 tons of fruits, vegetables, food, and flowers that can't be sold and would otherwise be thrown away. This organic material is placed in specially marked dumpsters at Larry's five stores. Coffee residuals, called "chaff," are used to control moisture and reduce odors in the dumpsters. Once a week, Lawson's Disposal, a local hauler, picks up the materials and transports them to Iddings, Inc., a nearby topsoil company, for composting. The materials are mixed with soil, yard trimmings, and other organic material to make a rich mixture that is sold for use as topsoil. The entire composting process takes 3 to 5 months to complete. Recently, Larry's has begun to buy back this mixture for use on company landscaping projects, thereby "closing the recycling loop." As a result of this composting project, Larry's Markets has reduced the amount of materials being landfilled by nearly 40 percent. This project has also significantly cut Larry's disposal costs. Composting a ton of material costs Larry's $67, while running compactors, hauling material to local landfills, and paying landfill fees and taxes costs $100 per ton. The difference between composting and landfilling for Larry's is a total savings of approximately $15,000 each year.

MSW with sewage biosolids, which is a form of co-composting. Co-composting is not discussed in detail in this guidebook.

During the composting process, feedstock is placed in a pile or windrow (an elongated pile) where decomposition takes place. The rate of decomposition depends on the level of technology used as well as on such physical, chemical, and biological factors as microorganisms, oxygen levels, moisture content, and temperature. Composting works best when these factors are carefully monitored and controlled.

The end products of a well-run composting process are a humus-like material, heat, water, and carbon dioxide. Compost is used primarily as a soil amendment or mulch by farmers, horticulturists, landscapers, nurseries, public agencies, and residents to enhance the texture and appearance of soil, increase soil fertility, improve soil structure and aeration, increase the ability of the soil to retain water and nutrients and moderate soil temperature, reduce erosion, and suppress weed growth and plant disease. Figures I-1 and I-2 at the end of this introduction illustrate the steps involved in composting yard trimmings and MSW.

Status of Composting Yard Trimmings and MSW in the United States

Nationwide, nearly 35 million tons of yard trimmings were generated in 1990, accounting for nearly 18 percent of the MSW stream (EPA, 1992). About 2,200 facilities for the composting of yard trimmings were operating in the United States in 1991 (Goldstein and Glenn, 1992). Approximately 12 percent or 4.2 million tons of the yard trimmings generated in 1990 were composted by these facilities (This estimate, however, does not include the amount of yard trimmings composted through "backyard" composting projects and other individual efforts.) (EPA, 1993).

In 1990, the United States also generated over 16 million tons of food scraps, 12 million tons of scrap wood, and 73 million tons of paper waste, which together account for 51 percent of the MSW stream (EPA, 1992). Although over 28 percent of all paper waste was recycled in 1988, a negligible amount of this material is currently composted (EPA, 1992). While composting of MSW has been practiced in other countries for many years, interest and commitment to MSW composting on a large scale is a recent development. As of 1992, 21 full-scale MSW composting facilities were in operation in the United States (Goldstein and Steuteville, 1992). Capacities of most of these facilities range from 10 to 500 tons of MSW feedstock per day. Minnesota leads the way, with eight operational facilities; Florida has three, and Wisconsin maintains two MSW composting facilities (see Table I-1). Minnesota's leading position is due, in part, to available state funds and technical assistance for MSW composting systems (Crawford, 1990). A number of facilities also are in the planning or construction stages (see Table I-2). Table I-3 provides a brief comparison of the composting of yard trimmings and MSW in reference to several operational and program parameters.

As these numbers indicate, composting is currently receiving a substantial amount of attention. Among other factors, this interest is due to regulatory and economic factors. In recent years, a number of communities and states have banned yard trimmings from disposal in landfills. As mentioned earlier, some states also have established ambitious landfill diversion goals, along with financial assistance programs that support alternative management projects. Several states also have adopted MSW compost regulations and more states are likely to follow. Another important legislative development is that several states currently require state agencies to purchase and use compost if it is available and if it is equivalent in quality to other soil amendments (Crawford, 1990).

Another indication of the headway being made in composting is the increasing number of vendors marketing their composting systems to public officials, haulers, and landfill operators (Goldstein and Glenn, 1992). In addition, many companies that are in the process of constructing new waste management facilities are planning to incorporate composting into their operations to reduce the amount of residuals that must be landfilled (Goldstein and Glenn, 1992). Additionally, many communities and commercial establishments are now attempting to compost a larger portion of the MSW stream in an effort to reuse materials, rather than landfill or combust them. Several municipalities have established pilot or ongoing programs to collect mixed MSW for composting. Others are conducting pilot projects for collecting source-separated food scraps. In addition, many restaurants and grocers are composting leftover or unusable food scraps at their operations.

Introduction Resources

Crawford, S. 1990. Solid waste/sludge composting: International perspectives and U.S. opportunities. Proceedings of the sixth international conference on solid waste management and secondary materials. Philadelphia, PA. December 4-7.

Glenn, J. 1992. The state of garbage in America: Part I. BioCycle. April, 33(4):46-55.

Goldstein, N., and J. Glenn. 1992. Solid waste composting plants face the challenges. BioCycle. November, 33(11):48-52.

Goldstein, N., and R. Steuteville. 1992. Solid waste composting in the United States. BioCycle. November, 33(11):44-47.

METRO. 1989. The art of composting: A community recycling handbook. Portland, Oregon: Metropolitan Service District.

Taylor, A., and R. Kashmanian. 1989. Yard Waste Composting: A Study of Eight Programs. EPA/530-SW-89-038. Washington, DC: Office of Policy, Planning and Evaluation and Office of Solid Waste and Emergency Response.

U.S. Environmental Protection Agency (EPA). 1992. U.S. Environmental Protection Agency. Characterization of Municipal Solid Waste in the United States. EPA/530-R-019. Washington, DC: Office of Solid Waste and Emergency Response.

U.S. Environmental Protection Agency (EPA). 1993. U.S. Environmental Protection Agency. Markets for compost. Washington, DC: Office of Solid Waste and Emergency Response and Office of Policy, Planning and Evaluation.

Introduction

Table I-1. Summary of operating MSW plants.

Plant Name	Year Started	Current Amount of MSW Composted (tons/day)	Proprietary Technology or System(1)	Ownership/Operation
Lakeside, AZ	1991	10-12	Bedminster Bioconversion	Joint Venture
New Castle, DE	1984	200-225	Fairfield digesters	Public/Private
Escambia County, FL	1991	200	—	Public/Public
Pembroke Pines, FL	1991	550	Buhler	Private/Private
Sumter County, FL	1988	50	—	Public/Private
Buenavista County, IA	1991	4,000/yr.	Lundell (for processing)	Private/Private
Coffeyville, KS	1991	50	—	Private/Private
Mackinac Island, MI	1992	8 (inc. MSW, sludge, manure) (2)	—	Public/Public
Fillmore County, MN	1987	12	—	Public/Public
Mora, MN (East Central SWC)	1991	250	Daneco	Public/Private
Lake of the Woods County, MN	1989	5	—	Public/Public
Pennington County, MN	1987	12	Lundell (for processing)	Public/Private
St. Cloud, MN	1988	60	Eweson digester w/Royer ag. bed	Private/Private
Swift County, MN	1990	12	—	Public/Public
Truman, MN (Prairieland SWB)	1991	55	OTVD	Public/Public
Wright County, MN	1992	165	Buhler	Public/Private
Sevierville, TN	1992	150 (design)	Bedminster Bioconversion	Public/Private
Hidalgo County, TX	1991	70	—	Public/Public
Ferndale, WA	1991	100	Royer ag. bed	Private/Private
Columbia County, WI	1992	40-45	—	Public/Public
Portage, WI	1986	20	—	Public/Public

(1) This category is limited to compost system vendors and not other proprietary technologies/equipment in use at these facilities.
(2) Amount for Mackinac Island indicates average daily flow due to peak population during the summer months.

Source: Goldstein and Glenn, 1992.

Table I-2. Nationwide listing of MSW composting facilities.

Facility	Status	System	Tons/Day (Unless noted)
ARIZONA			
1. Pinetop-Lakeside	Operational	Digester (Bedminster)	12-14 (w/6 wet tpd sludge)
ARKANSAS			
1. Madison	Pilot	Windrow	40 (100-150 at full scale)
CALIFORNIA			
1. Chowchilla (Madera County)	Planning	Windrow (enclosed)	500-800 (w/sludge)
2. San Diego (city)	Vendor negotiation	A-SP (Daneco)	300,000/year
3. Tulare County	Proposal review	Windrow or in-vessel	900-1,000 (w/sludge)
4. Ventura County	Proposal review for solid waste management		3,000 (total stream)
CONNECTICUT			
1. Northeastern Conn. Res. Rec. Auth. (Brooklyn)	Proposal review	In-Vessel/Enclosed	200
DELAWARE			
1. Del. Reclamation Project (New Castle)	Operational	In-Vessel (Fairfield)	200-225 (w/150-200 wet tpd sludge)
FLORIDA			
1. Cape Coral	Vendor negotiation	Windrow (Amerecycle)	200
2. Charlotte County	Proposal review		400 (w/sludge)
3. Escambia County	Operational	Windrow	200 (400 design)
4. Manatee County	Vendor negotiation	Windrow (Amerecycle)	1,000 (total stream)
5. Monroe County	Proposal review		75,600/yr. (total stream)
6. Palm Beach County	Pilot planned (1) (RDF, RDF rejects, mixed paper)	Agitated bed (IPS)	
7. Pembroke Pines	Operational	Enc. aerated windrow (Buhler)	550 (650 design)
8. Sumter County	Operational	Windrow (Amerecycle)	50
IOWA			
1. Buenavista County	Operational	Windrow (w/ Lundell processing line)	4,000/yr.
2. Cedar Rapids	Feasibility study (for wet/dry separation)		
3. Council Bluffs	Consideration		75-80
4. Harden County (w/ Butler, Wright Counties)	Planning	Windrow (w/ Lundell processing line)	60
KANSAS			
1. Coffeyville	Operational	Windrow	50
LOUISIANA			
1. Tri-Parrish SWC (St. Martin, Iberia, Lafayette)	Consideration of pilot (Cocomposting)		700 (total stream)
MAINE			
1. Bowdoinham	Pilot (Source sep. organics)	A-SP	1/month
2. Machias	Design (Source sep. organics)	Windrow	2,500 pop.
MARYLAND			
1. Baltimore	Construction (FERST Co.)	In-Vessel (A-S-H) (2)	520 (700 design)
2. Brandywine	Design/Financing (P.G. FERST)	Enc. A-SP (Rader Co.)	340
3. Salisbury (Wicomico Cty. Landfill)	Consideration of pilot (by American Materials Recyc.)	In-Vessel (Seerdrum)	20 (300 design)

Table I-2. (Continued).

Facility	Status	System	Tons/Day (Unless noted)
MASSACHUSETTS			
1. Berkshire County (Southern)	Consideration (Market/odor control studies needed)		15
2. Franklin County	Siting (for source sep. organics; sludge)	In-Vessel	100
3. Nantucket	Proposal review	Enc. aerated windrow	100
4. Northampton	Consideration		
5. Northfield	Proposed by Bennett Construction, Inc.	Tunnel w/ enc. windrows	200
6. Somerset	Proposed by ERS		700
7. Wrentham	Proposed by ERS		800 (w/ 180 tpd sludge)
MICHIGAN			
1. Mackinac Island	Operational	A-SP	8 (w/sludge, manure)
MINNESOTA			
1. Fillmore County	Operational	A-SP; Enc. windrow (in construction)	12
2. Freeborn/Mower Counties	Planning (Source sep. organics)	Aerated Windrow	80
3. Goodhue	Proposed by ERS		450
4. Kandiyohi County	Planning		20-40
5. Lake of the Woods County	Operational	Windrow	5
6. Mora (East Central SWC)	Operational	Enc. A-SP (Daneco)	250
7. Pennington County	Operational	Windrow	40 (80 tpd design)
8. Rice County	Planning (w/ addtl. counties)	Aerated Windrow	200
9. Rosemount	Planning (by Cares for RDF residuals, MSW, other organics)	Windrow	100
10. St. Cloud	Operational (by Recomp)	Digester w/ agitated bed	60 (100 tpd design)
11. St. Louis County	Consideration	A-SP	250
12. Swift County	Operational	Aerated Windrow	12 (5.5 to composting)
13. Truman (Prairieland Solid Waste Board)	Operational	In-Vessel (OTVD)	55 (100 design)
14. Wright County	Operational (RDF residuals from Anoka County)	Enc. aerated windrow (Buhler)	165
MISSOURI			
1. Springfield	Permitting	Enc. A-SP (Daneco)	500
NEW HAMPSHIRE			
1. Ashuelot Valley Refuse Disp. Dist.	Consideration		--
2. Hookset	Proposed by Aware Corp.	Windrow	800
NEW JERSEY			
1. Atlantic County	Design (Source sep. organics)	In-Vessel (agitated bed)	159 (w/sludge)
2. Cape May County	Permitting	Enc. A-SP (Daneco)	600
3. Ocean County	Pilot (Source sep. organics by Ocean Cty. Landfill Corp.)	In-Vessel	300 (design)
4. Ocean Township	Proposal review		400-500
NEW YORK			
1. Camillus	Planning	In-Vessel	50-100
2. Delaware County	Consideration	In-Vessel	100 (design)
3. East Hampton	Feasibility Study (Source sep. organics; MRF under construction)	In-Vessel	110 (total stream)
4. Eastern Rennselaer County SWMA	Vendor negotiation	Enc. A-SP (Daneco)	100-150 (w/sludge, septage)
5. Madison County	Pilot (Residential source sep. organics)	Windrow	2/wk.
6. Monroe County	Consideration	Windrow	300

Table I-2. (Continued).

Facility	Status	System	Tons/Day (Unless noted)
7. New York City	Pilot (Residential source sep. organics)	Windrow	4.5-5/wk.
8. Riverhead	Permitted by Omni Tech Serv. (Procuring waste contracts)	Agitated bed (Koch)	250 (500 design)
9. South Hampton	Consideration (Voter ref. on 11/3/92)		
NORTH CAROLINA			
1. Buncombe County	Pilot	Windrow	200 (total for pilot) 300-350 (design)
OREGON			
1. Portland	Temporarily closed (Vendor negotiation)		600 (design)
PENNSYLVANIA			
1. Adams County	Feasibility Study	In-Vessel	150 (total stream)
2. Blair County	Consideration		240 (total stream)
TENNESSEE			
1. Sevierville (Sevier Solid Waste)	Operational	Digester w/ aerated windrows (Bedminster)	150 (w/ 75 wet tpd sludge)
TEXAS			
1. Big Sandy	Operational by Bedminster (research facility)	Digester w/ windrow	35 w/ sludge (seasonal)
2. Hidalgo County	Operational	Windrow	70 (300 design)
3. Houston	Proposal by WPF Corp.	Aerated windrow	3,800/wk (total stream)
4. Stephenville	Permitting	Windrow	368,500 cy/yr.
VERMONT			
1. Central Vermont SWMD	Consideration		
VIRGINIA			
1. Loudon County	Planning (Source sep. organics by priv. co.)	Enc. aerated windrow	125-180
WASHINGTON			
1. Ferndale	Operational by Recomp	Digester w/agitated bed (Royer)	100
WEST VIRGINIA			
1. Leestown	Pilot (Source sep. organics)	Static pile, windrows	
WISCONSIN			
1. Adams County	Consideration (wet/dry)	Windrow	20
2. Burnett & Washburn Counties	Consideration	Windrow & A-SP	50
3. Columbia County	Operational	Drum w/enc. curing	40-45 w/ sludge (80 design)
4. Portage	Operational	Drum w/windrows	20 (w/sludge)
5. Vilas County	Pilot (in planning)	A-SP	30-50 for pilot 20 tpd (total stream)

Source: Goldstein and Glenn, 1992.

Table I-3. A brief comparison of yard trimmings and MSW composting.

Parameter	Yard Trimmings Composting	MSW Composting
Planning	Yard trimmings composting often requires planning for seasonal variations in the flow of feedstock.	Large-scale MSW composting will require a detailed waste stream assessment that will require planning and resources to complete.
Separation	Easy to separate yard trimmings feedstock from the rest of the MSW stream for collection and composting since yard trimmings are normally gathered by the homeowner separately from other materials.	Feedstock for MSW composting can be separated by residents or at the facility into recyclable, compostable, and/or noncompostable components.
Technology	Yard trimmings composting can be done using relatively simple technologies.	MSW composting requires more complex technology because it processes a mixed feedstock that can include varied contaminants.
Leachate Control	Leachate collection systems might be required, particularly for larger facilities and those in areas of moderate to high rainfall.	Due to the diversity of materials in MSW feedstock, leachate collection systems are generally required.
Odor Control	Yard trimmings compost facilities can often employ simple siting, process, and design controls to minimize odors.	MSW composting facilities are likely to require sophisticated technologies to control odors. More stringent siting and design measures also are typically needed.
Regulations	Yard trimmings composting is not governed by stringent regulations.	MSW composting is more stringently regulated or controlled than yard trimmings composting, and may require compliance with state or local permitting procedures.
Product Quality	Medium- to high-quality compost can be produced using relatively simple technology and can be easily marketed to end users.	Extensive preprocessing is required to achieve medium- to high-quality compost that can overcome public perception problems of impurity and be marketed easily.
Economics	A low-technology yard trimmings composting facility can be financed with a relatively small capital investment and low operating costs (mostly labor).	Siting, equipment, and permitting costs can add up to a large initial and ongoing investment for a MSW composting facility, particularly for a large operation.

Introduction

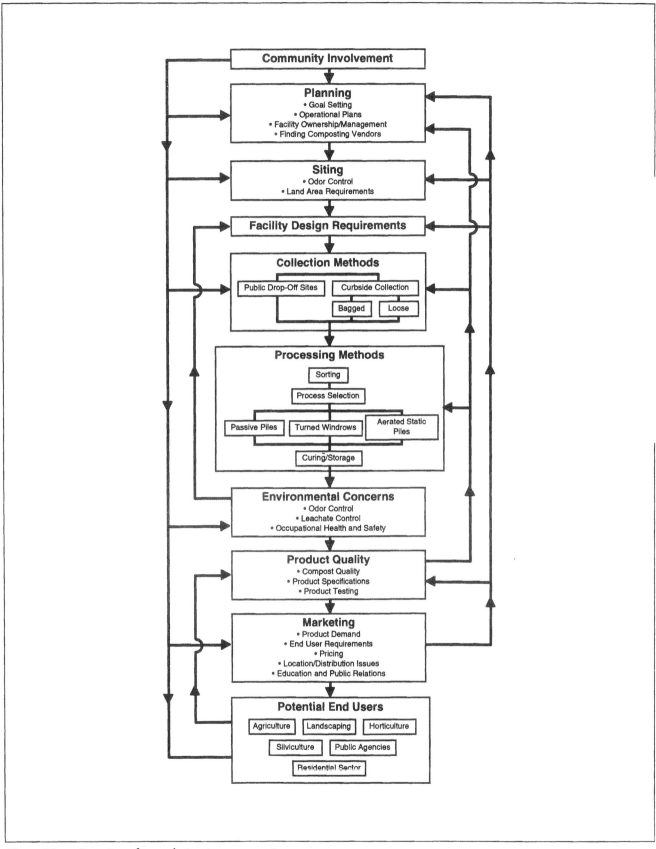

Figure I-1. Overview of a yard trimmings composting program.

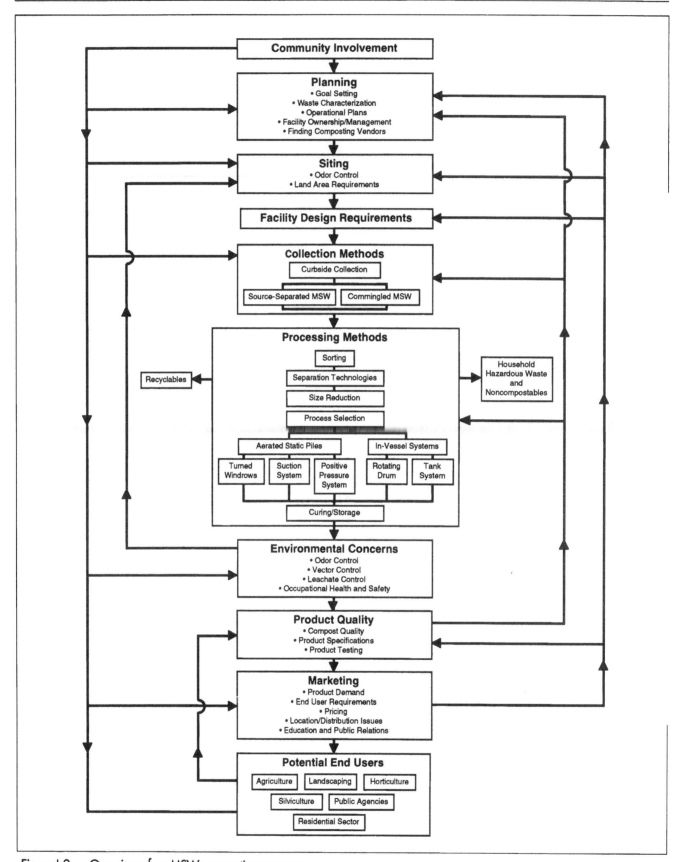

Figure I-2. Overview of an MSW composting program.

Chapter One
Planning

Communities that are considering incorporating composting into their solid waste management strategy need to conduct thorough planning to decide what type of program best fits the needs and characteristics of their locality. Because each community possesses its own set of financial, climactic, socio-economic, demographic, and land use characteristics, there is no formula dictating how to incorporate composting into an integrated waste management plan; these issues must be decided on a case-by-case basis for each community or region. This chapter describes some of the preliminary steps that a community should take before embarking on any composting program. Additional planning requirements are addressed throughout the guidebook.

Goal Setting

An important first step for public officials considering a composting program is to determine what they want the program to achieve. Typical goals of a composting program include:

- Reducing the flow of materials into landfills or combustors.
- Diverting certain types of materials from the MSW stream.
- Complying with state or local regulations or recovery goals.
- Providing a practical management option for a single community or a larger region.

Once a community has clearly defined the goals of its program, it will be easier to evaluate available technologies and determine the role that composting will play in the community's overall management strategy. In addition to goal-setting, it is important to evaluate the economic and technical feasibility of composting in the context of other waste management techniques, such as landfilling and combustion, to determine which alternatives are most suitable for the community. The costs and benefits of each option as well as relevant political and public opinion considerations can be evaluated to ascertain which mix of solid waste management approaches will best serve the community.

Waste Characterization

A municipal composting program must be implemented with a full understanding of the MSW stream. Identifying and quantifying the components of the local MSW stream should be an integral part of preliminary planning for every program. One way to obtain this information is to conduct a waste stream characterization study. These studies range in price from $35,000 to $400,000, depending on the type and quality of information needed. A comprehensive waste characterization study involves analyzing the local MSW stream by separating and sampling waste. Sampling can take place at the local waste management facility or at a transfer station. If a large-scale MSW composting facility is being contemplated, a detailed waste stream characterization study is necessary to ensure proper design (this would not be necessary in advance of a large-scale yard trimmings composting program). Publications, including the *Solid Waste Composition Study, 1990-1991: Part 1* published by the Minnesota Pollution Control Agency, are excellent references for more detailed information on conducting MSW stream assessments (this document is cited in the resources list at the end of this chapter). While a waste stream characterization study can provide information on the anticipated quantity of materials generated, it will not necessarily discern the amount of materials that will actually be collected or dropped off in the composting program since that will depend on factors such as the percent of homes or facilities that provide organic material for composting.

Although a comprehensive waste characterization study is the most accurate way to obtain data on the local MSW stream, the analysis involved can be very expensive and time consuming. Therefore, many communities might simply want to examine state or national MSW generation patterns, using these figures as a basis for determining local waste flow and characterization. Planners should, however, take into consideration any local factors that could influence the composition and amount of their MSW stream including:

- *Season and climate* - In certain parts of the country, the amount and type of yard trimmings generated will vary dramatically from season to season and as the climate changes. For example, an abundance of leaves are generated in autumn in many localities. Climate also can affect the composition and amount of the MSW stream. During warm seasons, for example, the quantity of beverage containers might be expected to rise. During the December holiday season, municipalities might expect a large amount of gift-wrapping paper or Christmas trees.

- *Regional differences* - Communities in Florida, for example, might discover that palm fronds constitute a large amount of their local MSW stream, while municipalities along the Maine seacoast must take into account large amounts of fish scraps generated in their region.

- *Demographics* - Population variations can have a significant impact on the MSW stream. These include temporary population changes (particularly in popular tourist or seasonal resort areas and college towns); the average age, income, and education of the population; age of neighborhoods; and population densities.

- *State of the economy* - The economic state of an area also can affect the composition of the MSW stream. For example, the increase in consumption that can be associated with good economic times might be reflected in an increase in packaging and other goods in the MSW stream.

- *Local source reduction and recycling programs* - Programs that aim to reduce or divert certain components of the MSW stream from disposal can affect the amount and type of materials that can be collected for composting.

For more accurate estimates, information from communities with similar demographic characteristics and sources of discards can be extrapolated to fit the local scene. Local collection services and facility operators also can be consulted. These individuals might have written records of the amount and type of discards collected on a yearly or even a monthly basis.

Operational Plans

An operational plan should be drafted to assist local officials and community members in understanding the proposed composting program and their roles in that program. An operational plan can be used as the basis for community discussion about the proposed program and for developing strong political support and consensus. The operational plan will be the community's road map for implementing and operating a successful composting program. Therefore, the more detailed the plan, the more useful it will likely be. The operational plan can be revised throughout the planning process as necessary to reflect major changes or alterations.

The operational plan should stipulate the chosen composting technology (e.g., turned windrows, aerated static piles, in-vessel systems, etc.); the equipment needed; proposed site design; and the pollution, nuisance, and odor control methods that will be employed. In addition, it should specify the personnel that will be required to operate the program as well as the type and extent of training they will require. The plan also should contain procedures for marketing or otherwise distributing the compost product.

When developing a plan, it is important to remember that all of the elements of a composting program (e.g., buying equipment, siting a facility, marketing the finished product, etc.) are interrelated. For this reason, all elements of a composting program should be chosen with other elements in mind. For example, composting site design can be influenced by a variety of factors. Site design might be influenced by the type of material that the site will process. A site which processes large quantities of a readily putrescible material and has close neighbors can require an enclosed design. Site design might also be influenced by compost markets. A site with screening capabilities and flexible retention time could be needed to meet the demands of end users. In addition, site design might be influenced by long-term considerations. A site with the potential to expand can be more appropriate for the community that expects its materials stream to grow in volume. As this example makes clear, decision-makers should accommodate the interrelated nature of the elements of a composting program throughout the planning process.

Community Involvement

Throughout the planning process, officials should work closely with collectors, haulers, processors, the recycling industry, local utilities, private citizens, and others to develop a safe, efficient, and cost-effective program. Providing these groups with a forum to express their concerns and ideas about composting will build a sense of ownership in the project as a whole. In addition, cooperation will enhance the understanding of the concerned groups about the compromises needed to make the program work; as a result, objections to siting or collection programs, for example, should be lessened. These groups also can provide invaluable information on vital aspects of a composting operation (see Chapter 10 for more information on community involvement).

Facility Ownership and Management

One of the basic decisions that must be addressed in the early planning stages is composting facility ownership and management. There are essentially four options for site ownership and operations, as shown in Table 1-1. These are municipal facilities, merchant facilities, privatized facilities, and contract services.

The option chosen for ownership and management of the composting facility will depend on many factors such as feedstock supply, land size and location, personnel resources, experience, costs, liability, financing methods, and political concerns. Composting facilities can be located on municipally or privately owned land, for example. When a community has available land and resources, it might consider owning and operating the facility itself. If the municipality has the land but not the resources for operation, it could contract out to an independent management firm. Communities might also consider encouraging the development of a privately owned and operated facility that works on a long-term contract, with the municipality guaranteeing tipping fees and feedstock. This facility might be owned and operated by a landfill owner or a refuse hauler that could serve the needs of all the communities it services. For larger facilities, in

Table 1-1. A comparison of facility ownership options.

Facility Type	Owner	Operator	Arrangement	Advantages	Disadvantages
Municipal	Municipality	Municipality	Municipality appoints a site manager, staffs the facility, and provides its own equipment.	Municipality has full control of operations.	Municipality shoulders all financial and performance risks associated with starting and operating the facility. If problems occur with the facility (e.g., traffic, odor, etc.), the municipality might have to address political issues as well.
Privatized	Private vendor	Private vendor	Vendor works under long-term service agreement with municipality to compost feedstock. Vendor designs and constructs facility on the basis of private capital attracted by the predictable revenue stream created by the long-term contract.	Municipality uses franchises and operating licenses to minimize competition for the vendor and thereby minimize investment risk for the vendor.	Municipality does not have full control over operations.
Merchant facility	Private vendor	Private vendor	Private vendor designs, finances, constructs, and operates facility on expectation of sufficient revenue from tipping fees and service charges. No contract between vendor and municipality exists, however.	Municipality carries no financial or operational risk.	High risk to vendor because of absence of contract guaranteeing feedstock and tipping fees. The public risk is tied to the possibility of the vendor failing and leaving the community with reduced waste management capacity. Also, community has no input on the level of service and no control of costs.
Contract services	Municipality	Private firm	Long-term contract with community for operation and maintenance of facility. Private company receives tipping fee. Municipality might staff the site or the private company might bring its own labor resources.	Municipality retains significant control since it can change service company upon expiration of the contract.	Municipality shoulders funding of facility.

Source: Gehr and Brown, 1992.

particular, municipalities should consider regional approaches to ownership and management. For example, one town might supply the site with others providing equipment and staffing. Such approaches offer both large and small communities advantages in financing, management, marketing, and environmental protection. Regional approaches also can help communities accomplish together what they cannot attain alone.

Composting Vendors

Many communities do not have the technical personnel and resources to develop and design a composting program and facility. It is not uncommon therefore to solicit this expertise from the private sector through a Request For Proposals (RFP). The purpose of an RFP is to encourage the submission of proposals from vendors that can conduct composting operations for the community. A well thought out and carefully worded RFP should include the broad operational plan for the community's composting program. This will give potential vendors the proper frame of reference for proposal development. In addition, the RFP should encourage the vendors to develop creative as well as low-cost options for composting. Finally, the RFP must provide a strong basis for reviewers to evaluate the different proposals and choose the vendor that offers the best mix of technical expertise, program design, and cost effectiveness for the community (Finstein et al., 1989).

Officials should consider hiring outside services to perform meticulous technical and economic analyses of any RFPs to determine their suitability to the community's specific solid waste characteristics. Given the plethora of source reduction, recycling, composting, and disposal options, many experts recommend the use of an RFP, particularly for more complex composting operations, in order to identify opportunities to maximize cost effectiveness and ensure the resulting composting operation will meet its goals.

Pilot Programs

Before implementing a full-fledged composting program, many communities first conduct pilot programs to determine the costs and prospects for success of a full-scale project. Pilot programs enable communities to experiment with different components of a program (such as composting technologies, collection strategies, and marketing techniques) to ascertain the most effective approaches for the community. Start-up costs for a pilot program are greater than for an ongoing composting program, however, and should not be used to estimate the start-up costs of a full-scale or long-term program.

Pilot Program in Seattle, Washington

From 1980 until 1989, the City of Seattle, Washington, conducted a yard trimmings composting pilot program consisting of a variety of demonstration projects aimed at determining the success of composting as a management option. Demonstration projects included community education on composting, Christmas tree recycling, and a 3-month "Clean Green" drop-off program for yard trimmings at the city's two transfer stations. In October 1988, Seattle passed an ordinance requiring residents to separate yard trimmings from recyclables and refuse. Based on the results of the city's pilot program, today Seattle maintains a three-pronged composting program: "Clean Green" drop-off centers for yard trimmings, backyard composting, and curbside collection of yard trimmings (ILSR, 1992).

Summary

In order to ensure a successful composting program, communities must plan ahead. Thorough planning will enable communities to detect any major problems with a composting operation that could jeopardize its success, such as an unacceptable siting decision, a lack of consistent feedstock, or a shortage of demand for the final product. Among the preliminary planning steps that a community should undertake are setting goals, conducting a waste stream characterization study or assessment, developing an operational plan, soliciting the viewpoints of affected parties, determining site ownership and management, securing a vendor, and considering the value of conducting a pilot program. Officials should view composting as one alternative in their MSW management program and analyze its effectiveness in comparison with management alternatives including source reduction, landfilling, and combustion.

Chapter One Resources

Finstein, M., P. Strom, F. Miller, and J. Hogan. 1990. Elements of a request for proposal (RFP) for sludge and municipal solid waste composting facilities: Scientific and technical aspects. New Brunswick, NJ: Rutgers Cooperative Extension, New Jersey Agricultural Experiment Station.

Gehr, W., and M. Brown. 1992. When privatization makes sense. BioCycle. July, 33(7): 66-69.

International City/County Management Association (ICMA). 1992. Composting: Solutions for waste management. Washington, DC: ICMA

Institute for Local Self-Reliance (ILSR). 1992. In-depth studies of recycling and composting programs: Designs, costs, results. Volume III.

Minnesota Pollution Control Agency (MPCA). 1992. Solid Waste Composition Study, 1990-1991: Part 1. St. Paul, MN: Ground Water and Solid Waste Division.

O'Leary, P., P. Walsh, and A. Razvi. 1990. Composting and the waste management plan. Waste Age. February, 21(2): 112-117.

U.S. Environmental Protection Agency (EPA). 1989. Decision-Maker's Guide to Solid Waste Management. EPA/530-SW-89-072. Washington, DC: Office of Solid Waste and Emergency Response.

Walsh, P., A. Razvi, and P. O'Leary. 1990. Operating a successful compost facility. Waste Age. January, 21(1): 100-106.

Chapter Two
Basic Composting Principles

Composting relies on a natural process that results from the decomposition of organic matter by microorganisms. Decomposition occurs wherever organic matter is provided with air and moisture; it occurs naturally on the forest floor and in open fields. Composting, as the term is used in this guidebook, is distinguished from this kind of natural decomposition in that certain conditions or parameters (such as temperature and moisture) are "controlled" to optimize the decomposition process and to produce a final product that is sufficiently stable for storage and application to land without adverse environmental impacts. This chapter provides a brief introduction to the biology involved in composting. It also describes the physical and chemical parameters that influence the process. Chapter 4 of this guidebook discusses how to control these parameters to optimize composting.

Overview of the Composting Process

The composting process occurs in two major phases. In the first stage, microorganisms decompose the composting feedstock into simpler compounds, producing heat as a result of their metabolic activities. The size of the composting pile is reduced during this stage. In the second stage, the compost product is "cured" or finished. Microorganisms deplete the supply of readily available nutrients in the compost, which, in turn, slows their activity. As a result, heat generation gradually diminishes and the compost becomes dry and crumbly in texture. When the curing stage is complete, the compost is considered "stabilized" or "mature." Any further microbial decomposition will occur very slowly.

The Role of Microorganisms

Composting is a succession of microbial activities whereby the environment created by one group of microorganisms invites the activity of successor groups. Different types of microorganisms are therefore active at different times in the composting pile. Bacteria have the most significant effect on the decomposition process, and are the first to take hold in the composting pile, processing readily decomposable nutrients (primarily proteins, carbohydrates, and sugars) faster than any other type of microorganism. Fungi, which compete with bacteria for food, play an important role later in the process as the pile dries, since fungi can tolerate low-moisture environments better than bacteria. Some types of fungi also have lower nitrogen requirements than bacteria and are therefore able to decompose cellulose materials, which bacteria cannot. Because fungi are active in composting piles, concern has arisen over the growth of opportunistic species, particularly those belonging to the genus *Aspergillus*. Chapter 6 discusses the potential health risks associated with this fungus.

Macroorganisms also play a role in the composting process. Rotifers, nematodes, mites, springtails, sowbugs, beetles, and earthworms reduce the size of the composting feedstock by foraging, moving in the compost pile, or chewing the composting materials. These actions physically break down the materials, creating greater surface area and sites for microbial action to occur.

The microorganisms necessary for composting are naturally present in most organic materials, including leaves, grass clippings, and other yard trimmings, and other organic materials. Products are available that claim to speed the composting process through the introduction of selected strains of bacteria, but tests have shown that inoculating compost piles in this manner is not necessary for effective composting of typical yard trimmings or MSW feedstock (Rynk et al., 1992; Haug, 1980; Gray et al., 1971a).

The bacteria and fungi important in decomposing the feedstock material can be classified as mesophilic or thermophilic. Mesophilic microorganisms or mesophiles (those that grow best at temperatures between 25 and 45°C [77 to 113°F]) are dominant throughout the

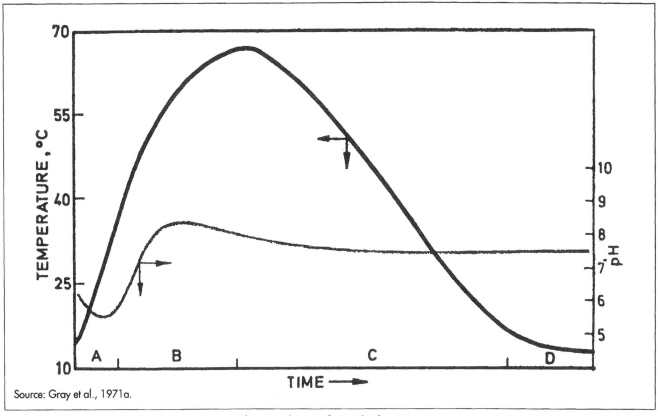

Figure 2-1. Temperature and pH variation with time: phases of microbial activity.
A = mesophilic, B = thermophilic, C = cooling, D = maturing.

composting mass in the initial phases of the process when temperatures are relatively low. These organisms use available oxygen to transform carbon from the composting feedstock to obtain energy, and, in so doing, produce carbon dioxide (CO_2) and water. Heat also is generated as the microorganisms metabolize the composting feedstock. As long as the compost pile is of sufficient size to insulate internal layers from ambient temperatures and no artificial aeration or turning occurs, most of the heat generated by the microorganisms will be trapped inside the pile. In the insulated center layers, temperatures of the composting mass will eventually rise above the tolerance levels of the mesophilic organisms. Figure 2-1 shows a typical temperature pattern for natural composting processes. When the temperatures reach toward 45°C (113°F), mesophiles die or become dormant, waiting for conditions to reverse.

At this time, thermophilic microorganisms or thermophiles (those that prefer temperatures between 45 and 70°C [113 and 158°F]) become active, consuming the materials readily available to them, multiplying rapidly, and replacing the mesophiles in most sections of the composting pile. Thermophiles generate even greater quantities of heat than mesophiles, and the temperatures reached during this time are hot enough to kill most pathogens and weed seeds. Many composting facilities maintain a temperature of 55°C (131°F) in the interior of the compost pile for 72 hours to ensure pathogen destruction and to render weeds inviable. (See Chapter 6 for a detailed discussion of pathogens and Chapter 7 for a discussion of different states' requirements for ensuring pathogen and weed destruction.)

The thermophiles continue decomposing the feedstock materials as long as nutrient and energy sources are plentiful. As these sources become depleted, however, thermophiles die and the temperature of the pile drops. Mesophiles then dominate the decomposition process once again until all readily available energy sources are utilized. Table 2-1 shows the density of microorganisms as a function of temperature during composting.

Factors Influencing the Composting Process

Because microorganisms are essential to composting, environmental conditions that maximize microbial activity will maximize the rate of composting. Microbial activity is influenced by oxygen levels, particle sizes of the feedstock material, nutrient levels and balance (indicated by the carbon-to-nitrogen ratio), moisture content, temperature, and acidity/alkalinity (pH). Any changes in these factors are interdependent; a change in one parameter can often result in changes in others. These factors and their interrelationships are discussed briefly below and in more detail in Chapter 4.

Oxygen

Composting can occur under aerobic (requires free oxygen) or anaerobic (without free oxygen) conditions, but aerobic composting is much faster (10 to 20 times faster) than anaerobic composting. Anaerobic composting also tends to generate more odors because gases such as hydrogen sulfide and amines are produced. Methane also is produced in the absence of oxygen.

Microorganisms important to the composting process require oxygen to break down the organic compounds in the composting feedstock. Without sufficient oxygen, these microorganisms will diminish, and anaerobic microorganisms will take their place. This occurs when the oxygen concentration in the air within the pile falls below 5 to 15 percent (ambient air contains 21 percent oxygen). To support aerobic microbial activity, void spaces must be present in the composting material. These voids need to be filled with air. Oxygen can be provided by mixing or turning the pile, or by using forced aeration systems (Chapter 4 discusses mixing and aeration methods in more detail).

The amount of oxygen that needs to be supplied during composting depends on:

- *The stage of the process* - Oxygen generally needs to be supplied in the initial stages of composting; it usually does not need to be provided during curing.
- *The type of feedstock* - Dense, nitrogen-rich materials (e.g., grass clippings) will require more oxygen.
- *The particle size of the feedstock* - Feedstock materials of small particle size (e.g., less than 1 or 2 inches in diameter) will compact, reducing void spaces and inhibiting the movement of oxygen. For this reason, the feedstock should not be shredded too small before processing (see below and Chapter 4 for information on size reduction).
- *The moisture content of the feedstock* - Materials with high moisture content (e.g., food scraps, garden trimmings) will require more oxygen.

Care must be taken, however, not to provide too much aeration, which can dry out the pile and impede composting.

Particle Size

The particle size of the feedstock affects the composting process. The size of feedstock materials entering the composting process can vary significantly. In general, the smaller the shreds of composting feedstock, the higher the composting rate. Smaller feedstock materials have greater surface areas in comparison to their volumes. This means that more of the particle surface is exposed to direct microbial action and decomposition in the initial stages of composting. Smaller particles within the composting pile also result in a more homogeneous mixture and improve insulation (Gray et al., 1971b). Increased insulation capacity helps maintain optimum temperatures in the composting pile. At the same time, however, the particles should not be so small as to compact too much, thus excluding oxygen from the void spaces, as discussed above. (Chapter 4 describes techniques for size reducing composting feedstock prior to processing.)

Table 2-1. Microbial populations during aerobic composting.[a]

Microbe	Number per Wet Gram of Compost			Numbers of Species Identified
	Mesophilic Initial Temp (40°C)	Thermophilic (40-70°C)	Mesophilic (70°C to Cooler)	
Bacteria				
Mesophilic	10^8	10^6	10^{11}	6
Thermophilic	10^4	10^9	10^7	1
Actinomycetes				
Thermophilic	10^4	10^8	10^5	14
Fungi[b]				
Mesophilic	10^6	10^3	10^5	18
Thermophilic	10^3	10^7	10^6	16

[a]Composting substrate not stated but thought to be garden-type materials composted with little mechanical agitation.
[b]Actual number present is equal to or less than the stated value.
Source: Haug, 1980.

Nutrient Levels and Balance

For composting to proceed efficiently, microorganisms require specific nutrients in an available form, adequate concentration, and proper ratio. The essential macronutrients needed by microorganisms in relatively large amounts include carbon (C), nitrogen (N), phosphorus (P), and potassium (K). Microorganisms require C as an energy source. They also need C and N to synthesize proteins, build cells, and reproduce. P and K are also essential for cell reproduction and metabolism. In a composting system, either C or N is usually the limiting factor for efficient decomposition (Richard, 1992a).

Composting organisms also need micronutrients, or trace elements, in minute amounts to foster the proper assimilation of all nutrients. The primary micronutrients needed include boron, calcium, chloride, cobalt, copper, iron, magnesium, manganese, molybdenum, selenium, sodium, and zinc (Boyd, 1984). While these nutrients are essential to life, micronutrients present in greater than minute amounts can be toxic to composting microorganisms.

Even if these nutrients are present in sufficient amounts, their chemical form might make them unavailable to some or all microorganisms. The ability to use the available organic compounds present depends on the microorganism's "enzymatic machinery" (Boyd, 1984). Some microorganisms cannot use certain forms of nutrients because they are unable to process them. Large molecules, especially those with different types of bonds, cannot be easily broken down by most microorganisms, and this slows the decomposition process significantly. As a result, some types of feedstock break down more slowly than others, regardless of composting conditions (Gray et al., 1971a). For example, lignin (found in wood) or chitin (present in shellfish exoskeletons) are very large, complex molecules and are not readily available to microorganisms as food. These materials therefore decompose slowly.

The C:N ratio is a common indicator of the availability of compounds for microbial use. The measure is related to the proportion of carbon and nitrogen in the microorganisms themselves. (Chapter 4 discusses the C and N content of different types of feedstock.)

High C:N ratios (i.e., high C and low N levels) inhibit the growth of microorganisms that degrade compost feedstock. Low C:N ratios (i.e., low C and high N levels) initially accelerate microbial growth and decomposition. With this acceleration, however, available oxygen is rapidly depleted and anaerobic, foul-smelling conditions result if the pile is not aerated properly. The excess N is released as ammonia gas. Extreme amounts of N in a composting mass can form enough ammonia to be toxic to the microbial population, further inhibiting the composting process (Gray et al., 1971b; Haug, 1980). Excess N can also be lost in leachate, in either nitrate, ammonia, or organic forms (Richard, 1992b) (see Chapter 6).

Moisture

The moisture content of a composting pile is interconnected with many other composting parameters, including moisture content of the feedstock (see Chapter 4), microbial activity within the pile, oxygen levels, and temperature. Microorganisms require moisture to assimilate nutrients, metabolize new cells, and reproduce. They also produce water as part of the decomposition process. If water is accumulated faster than it is eliminated via either aeration or evaporation (driven by high temperatures), then oxygen flow is impeded and anaerobic conditions result (Gray et al., 1971b). This usually occurs at a moisture level of about 65 percent (Rynk et al., 1992).

Water is the key ingredient that transports substances within the composting mass and makes the nutrients physically and chemically accessible to the microbes. If the moisture level drops below about 40 to 45 percent, the nutrients are no longer in an aqueous medium and easily available to the microorganisms. Their microbial activity decreases and the composting process slows. Below 20 percent moisture, very little microbial activity occurs (Haug, 1980).

Temperature

Temperature is a critical factor in determining the rate of decomposition that takes place in a composting pile. Composting temperatures largely depend on how the heat generated by the microorganisms is offset by the heat lost through controlled aeration, surface cooling, and moisture losses (Richard, 1992a) (see Chapter 4). The most effective composting temperatures are between 45 and 59°C (113 and 138°F) (Richard, 1992a). If temperatures are less than 20°C (68°F), the microbes do not proliferate and decomposition slows. If temperatures are greater than 59°C (138°F), some microorganisms are inhibited or killed, and the reduced diversity of organisms results in lower rates of decomposition (Finstein et al., 1986; Strom, 1985).

Microorganisms tend to decompose materials most efficiently at the higher ends of their tolerated temperature ranges. The rate of microbial decomposition therefore increases as temperatures rise until an absolute upper limit is reached. As a result, the most effective compost managing plan is to maintain temperatures at the highest level possible without inhibiting the rate of microbial decomposition (Richard, 1992a; Rynk et al., 1992).

Acidity/Alkalinity (pH)

The pH of a substance is a measure of its acidity or alkalinity (a function of the hydrogen ion concentration), described by a number ranging from 1 to 14. A pH of 7 indicates a neutral substance, whereas a substance with pH level below 7 is considered to be acidic, and a substance with a pH higher than 7 is alkaline. Bacteria prefer a pH between 6 and 7.5. Fungi thrive in a wider range of pH levels than bacteria, in general preferring a pH between 5.5 and 8 (Boyd, 1984). If the pH drops below 6, microorganisms, especially bacteria, die off and decomposition slows (Wiley, 1956). If the pH reaches 9, nitrogen is converted to ammonia and becomes unavailable to organisms (Rynk et al., 1992). This too slows the decomposition process.

Like temperature, pH levels tend to follow a successional pattern through the composting process. Figure 2-1, on page 17, shows the progression of pH over time in a composting pile. As is illustrated, most decomposition takes place between pH 5.5 and 9 (Rynk et al., 1992; Gray et al., 1971b). During the start of the composting process, organic acids typically are formed and the composting materials usually become acidic with a pH of about 5. At this point, the acid-tolerating fungi play a significant role in decomposition. Microorganisms soon break down the acids, however, and the pH levels gradually rise to a more neutral range, or even as high as 8.5. The role of bacteria in composting increases in predominance again as pH levels rise. If the pH does not rise, this could be an indication that the compost product is not fully matured or cured.

Summary

Composting is a biological process influenced by a variety of environmental factors, including the number and species of microorganisms present, oxygen levels, particle size of the composting materials, nutrient levels, moisture content, temperature, and pH. All of these factors are interrelated, and must be monitored and controlled throughout the composting process to ensure a quality product.

Chapter Two Resources

Alexander, M. 1961. Introduction to soil microbiology. New York, NY: Wiley Publishing Co. As cited in: Gray, K.R., K. Sherman, and A.J. Biddlestone, 1971b. A review of composting: Part 2 - The practical process. Process Biochemistry. 6(10):22-28.

Boyd, R.F. 1984. General microbiology. Wirtz, VA: Time Mirror/Mosby College Publishing.

Finstein, M.S., F.C. Miller, and P.F. Strom. 1986. Waste treatment composting as a controlled system. As cited in: H.J. Rehm and G. Reed, eds.; W. Schonborn, vol. ed. Biotechnology: A comprehensive treatise in 8 volumes: Volume 8, microbial degradations. Weiheim, Germany: Verlagsangabe Ver. Chemie (VCH).

Golueke, C.G. 1977. Biological reclamation of solid wastes. Emmaus, PA: Rodale Press.

Gray, K.R., K. Sherman, and A.J. Biddlestone. 1971a. A review of composting: Part 1 - Process Biochemistry. 6(6):32-36.

Gray, K.R., K. Sherman, and A.J. Biddlestone. 1971b. A review of composting: Part 2 - The practical process. Process Biochemistry. 6(10):22-28.

Haug, R.T. 1980. Compost engineering principles and practice. Ann Arbor, MI: Ann Arbor Science Publishers, Inc.

Massachusetts Department of Environmental Protection (MA DEP). 1991. Leaf and yard waste composting guidance document. Boston, MA: MA DEP, Division of Solid Waste Management.

May, J.H., and T.W. Simpson. Virginia Polytechnic Institute and State University. 1990. The Virginia yard waste management manual. Richmond, VA: Virginia Department of Waste Management.

McGaughy, P.H., and H.G. Gotaas. Stabilization of municipal refuse by composting. American Society of Civil Engineers. Paper No. 2767. As cited in: Haug, 1980. Compost engineering principles and practice. Ann Arbor, MI: Ann Arbor Science Publishers, Inc.

Poincelot, R.P. 1977. The biochemistry of composting. As cited in: Composting of Municipal Residues and Sludges: Proceedings of the 1977 National Conference. Rockville, MD: Information Transfer.

Richard, T.L. 1992a. Municipal solid waste composting: Physical and biological processing. Biomass & Bioenergy. Tarrytown, NY: Pergamon Press. 3(3-4):163-180.

Richard, T.L. 1992b. Personal communication. College of Agriculture and Life Sciences. Cornell University. Ithaca, NY.

Rynk, R., et al. 1992. On-farm composting handbook. Ithaca, NY: Cooperative Extension, Northeast Regional Agricultural Engineering Service.

Strom, P.F. 1985. Effect of temperature on bacterial species diversity in thermophilic solid waste composting. Applied Environmental Microbiology. 50(4): 899-905. As cited in: Richard, 1992a. Municipal solid waste composting: Physical and biological processing. Biomass & Bioenergy. Tarrytown, NY: Pergamon Press. 3(3-4):163-180.

Wiley, J.S. 1956. Proceedings of the 11th industrial waste conference. Purdue University, series 91, p. 334.

Chapter Three
Collection Methods

The cost, ease, and effectiveness of implementing a composting program is affected by the method chosen for collecting the compost feedstock. Communities can select from a variety of collection systems to develop a composting program to meet their specific needs. Programs can be designed to collect just yard trimmings, or yard trimmings and MSW. Collection can occur at curbside, where the municipality picks up the materials directly from households, or through drop-off sites, where residents and commercial producers deliver their compostable materials to a designated site. Most communities will want to build on their existing refuse collection infrastructure when implementing a composting program. This will ease the implementation of composting into a community's overall MSW management program and help contain costs. This chapter describes the advantages and disadvantages of various collection methods and examines some of the factors that decision-makers should consider when examining the applicability of different systems. Because collection is very different depending on whether yard trimmings, MSW, or both are being collected, this chapter is divided into two halves. The first portion of the chapter discusses yard trimmings collections; the second section focuses on source-separated and commingled MSW collections.

Factors in Yard Trimmings Collection

When developing a yard trimmings collection program, officials must take into account the length of the growing season, which affects both the amount of feedstock to be collected as well as the duration of collection. In the more temperate climates of the southern and southwestern regions of the United States, collection can take place throughout the year. In other areas of the United States, collecting yard trimmings is largely a seasonal matter.

Grass can be collected from spring through fall (the average growing season is 24 to 30 weeks). Leaves usually can be collected from mid-October through December and once again in the spring. Brush typically is collected in spring and fall. Depending on the season and the region, the brush, grass, and leaves can be collected together or separately. Ideally, brush should not be mixed with grass cuttings and leaves during collection without first being processed into smaller pieces because large branches tend to decompose more slowly. Because large volumes of leaves are generated within a relatively short time span, many communities find it cost-effective to collect and compost them separately from other yard trimmings. Leaves can be composted with other materials, usually grass, whose high nitrogen content can accelerate the composting process and result in a higher quality finished product (see Chapters 2 and 4). The high nitrogen content of grass can, however, cause odor problems during the composting process if not balanced with sufficient carbonaceous material and managed properly (see Chapters 4 and 6 for more information).

There are two basic options for collecting yard trimmings: public drop-off sites and curbside collection. When establishing a collection program, community leaders must consider the program's convenience for the public, as well as the level of interest displayed by citizens participating in the program. A drop-off program in a small, densely-populated community with residents well-educated about the importance of composting might garner high participation rates. By contrast, in a community that is uninterested or uneducated about composting, even a curbside program—which is typically more convenient for community residents—might fail to bring in large amounts of yard trimmings. Drop-off and curbside collection methods are described below.

Public Drop-Off Sites for Yard Trimmings

Public drop-off sites are specified locations where residents and businesses can take their yard trimmings. Drop-off sites can be an effective, low-cost option for some municipalities since they allow communities to operate a composting program while avoiding the labor and capital investment costs associated with curbside collection operations.

Home Composting and "Don't Bag It" Programs

Residents can be encouraged to let grass clippings remain on the lawn. The clippings will decompose and add nutrients to the soil. This eliminates the need to bag and remove the cuttings. Although exact recommendations depend on the variety of grass, it is generally advisable to not cut more than one-third of the blade, and not more than 1 inch total at any time. Leaves also can be mulched with a lawn mower into the lawn if cut finely enough.

Home composting of yard trimmings also serves to divert material from being collected and recovered or disposed of. Additionally, residents are provided with compost for gardening and landscaping. Home composting is particularly appropriate for residential lots of one-half acre or larger. Many types of food scraps can be composted as well.

To encourage individuals to leave clippings on the lawn, perform mulching, or compost at home, municipalities must educate residents about the "whys" and "hows" of these procedures. Many towns and cities, states, and university extension services across the country have published local guides and brochures on how to mulch and compost. Also, incentives such as providing simple compost bins at no cost can encourage residential composting.

Drop-off stations can be located at established recycling centers, landfills, and transfer stations or at the composting facility itself. In addition, some localities employ a system of collection trailers, which can travel to different locations in the community for added convenience to area residents. In all cases, yard trimmings should be collected frequently from drop-off centers to prevent the formation of odors and attraction of vectors (see Chapters 4 and 6).

Drop-off collections typically have low participation rates primarily because residents must assume the responsibility for collection (Richard et al., 1990). In communities where citizens are accustomed to delivering their household waste to landfills or transfer stations, drop-off collections of yard trimmings are more likely to succeed. Drop-off programs in communities with curbside collection of MSW, however, could witness lower collection rates at first due to residents' lack of familiarity with this collection method. To encourage participation, communities should strive to make the collection as convenient as possible. Some programs, for example, allow participants to pick up finished compost, firewood, or wood chips on the same day they drop off compostable materials. In addition, the public should be informed of the specifics of the community's collection program, as well as the rationale for and benefits of composting (see Chapter 10 for more information on community involvement).

In addition to residents, other sectors of the community can be encouraged to participate in yard trimmings drop-off programs. For example, businesses that generate a substantial amount of yard trimmings (such as landscape contractors) might be allowed to drop off the material. In areas where tipping fees are charged for municipal solid waste disposal, businesses might be offered a reduced fee as an incentive for bringing in yard trimmings for composting. This would mean, however, that incoming shipments would need to be measured. To eliminate the need for measuring shipments on site, communities could calculate the average amount of yard trimmings per truckload (based on tons or pounds per cubic yard) for each business and draft permits for a limited number of drop-offs based on these calculations. Figure 3-1 presents a sample yard trimmings drop-off permit.

Curbside Collection of Yard Trimmings

In a curbside collection program, the municipality picks up the yard trimmings that residents have placed outside of their homes. Curbside collection of yard trimmings typically offers the advantage of higher participation rates than drop-off programs. Overall, curbside collection is more expensive than drop-off collection due to the added equipment and labor resources needed. Nevertheless, additional costs are frequently justified by the volume of yard trimmings that is diverted and recovered.

The frequency of pickup will depend on such factors as the type and amount of yard trimmings being collected, the size and makeup of the community, and the budget. Schedules for curbside collection can range from weekly collections for grass in the summer to a single annual collection for brush.

Communities also must decide which collection method to employ for curbside yard trimmings collection. The material either can be collected in a container set out by the household or collected loose with the aid of a front loader or other equipment (see Appendix B). Several programs, such as those in Columbia, South Carolina, and Sacramento, California, have been collecting loose yard trimmings since the 1950s or earlier (Glenn, 1989). Collection of containerized yard trimmings, on the other hand, is relatively new. The advantages and disadvantages of both collection strategies are examined below.

Loose Yard Trimmings

Picking up loose yard trimmings at the curbside, a practice known as bulk collection, is most frequently used for collecting leaves during fall when communities generate large volumes of this material. Bulk collection avoids the cost of providing bags or special containers to residents

Permit #_____ _____ Last Name First	Mo. Day Yr. (__/__/__)	Disposal Quantity Purchased Cu. Yd.	Amount Paid ($) Cu. Yd.	Quantity Used	Unused Cu. Yd.

Name _____

Street _____ P.O. Box _____

City _____ Zip _____ Phone _____

<u>Classification of yard waste source (mark (x) one)</u>

( ) 1. individual residence   ( ) 5. commercial property
( ) 2. commercial gardener   ( ) 6. public utility
( ) 3. tree surgeon   ( ) 7. local government:_____
( ) 4. school/college   ( ) 8. other government unit:_____
     ( ) 9. other (specify) _____

—Leaf Disposal Information for Commercial Haulers—

- There will be a permit fee of $_____ per each vehicle for dumping at this site. The permit will be affixed on the inside of the window of the driver's side, and be in plain view upon entering the composting site.

- Permits may be obtained at the compost site or city hall, Monday through Friday from 9:00 AM to 11:00 AM only. Payment shall be a certified check or money order made out to the Town of _____. <u>NO CASH WILL BE ACCEPTED.</u>

- The hours of operation will be Monday through Friday from 7:00 AM to 5 PM beginning _____. There will be no dumping on Veteran's Day and Thanksgiving Day. Dumping will terminate on _____, or sooner, at the discretion of the public works superintendent, if the yard becomes full.

- Haulers depositing yard waste will enter and exit from_____.

- The DPW requests the cooperation of all permit holders and reminds everyone that no plastic bags or any other foreign materials are to be included with the yard waste. Failure to follow any of the above mentioned, or the instructions of the site attendant, may result in the forfeiture of one's permit.

- Permits are granted as an exclusive right of the DPW and are to be used only at the compost site. Said permits are non-transferrable and may be revoked for just cause at any time.

Source: Richard et al., 1990.

Figure 3-1. Yard trimmings drop-off permit application form from New York State.

Collection Methods

(or requiring residents to purchase these items). In addition, bulk collection facilitates the unloading of material at the facility since no debagging is necessary.

Bulk collections are a long, labor-intensive process, however, and could require the community to purchase new equipment. As a result, the community might be able to afford only a reduced pickup schedule. Many different types of equipment are used to pick up unbagged leaves mechanically. Vacuum trucks are commonly used to collect piles of leaves. These trucks often can mix leaves with glass, sand, and other undesirable substances found on the road, however, and are not effective when the leaves become wet or frozen (See Appendix B for more information). Front-end loaders can be used under these conditions but are not effective with dry leaves. Specialized vehicles, such as tractors equipped with a claw or leaf-loaders that quickly sweep material from the curb to the transportation truck, are becoming available for bulk collections of yard trimmings. (See Appendix B for descriptions and costs for specialized equipment.)

Communities must consider several potential problems inherent in bulk leaf collections. First, loose leaves are susceptible to being mixed with unwanted objects such as glass, cans, and car batteries (Richard et al., 1990). The leaves also become difficult to collect after they have blown around or children have played in them. In addition, loose leaves can catch fire from hot automobile exhaust systems.

Bulk collection of unbagged brush and grass clippings is problematic. Piles of grass left on the sidewalk are very difficult to collect, and in most communities this option is not cost-effective. Brush collections require special handling. Because brush does not readily compact, mobile wood chippers might be needed to reduce the volume of brush, thereby facilitating collection and cutting down on handling and transportation costs. Alternatively, brush can be collected in bundles and taken to a central processing facility for chipping. While brush is produced year round, it is impractical to have a year-round collection program because of the relatively small amount of material involved. Many communities have organized monthly or annual brush collection days (Mielke et al., 1989).

Bagged or Containerized Yard Trimmings

Collecting bagged or containerized yard trimmings at the curbside is typically a neater and more efficient operation than collecting in bulk. Moving the materials to the transportation vehicle is relatively quick, and the bags or containers are not affected seriously by weather conditions. Communities generally can use a standard compactor truck for collection. Furthermore, existing programs have found that bagged yard trimmings typically contain less noncompostable material than unbagged yard trimmings.

Several types of containers can be used for collection. Common containers include plastic and degradable plastic bags, paper bags, and specialized marked trash containers. Table 3-1 lists the major advantages and disadvantages of each type of bag and bin. Another alternative the

---

Collection Strategies in Two Massachusetts Towns

Melrose, Massachusetts, opened a leaf composting facility in October 1990 in response to a state landfill ban on leaves and other yard trimmings. To cover costs, the Boston suburb invited other regional communities to send leaves to the facility for a moderate tipping fee with one stipulation—that the leaves be delivered loose or in biodegradable paper bags.

Several towns and cities in the area responded immediately, including Stoneham and Burlington. Stoneham officials decided to collect leaves at the curbside throughout the entire town on two Saturdays at the beginning of November and December, respectively. Six biodegradable paper bags would be provided at no cost to each household, with extra bags available at the Stoneham Department of Public Works at the cost of three for $1. Stoneham also established a 40-cubic-yard container at the Department of Public Works where residents could drop off leaves from October 1 through December 15, 1990. Because of Stoneham's compact size—no household was located more than 5 minutes from the drop-off site—the combination of limited curbside collection and a drop-off container worked to capture about 60 percent of the estimated leaf stream available.

Burlington officials, on the other hand, decided against a drop-off center in favor of more frequent curbside collections. This was due primarily to the more dispersed population of the town. (A central drop-off location would make it inconvenient for some households to drive the 20 minutes necessary to deposit their leaves.) Burlington officials contracted with the town collector to pick up all available leaves for 6 weeks in the fall and 3 weeks in the spring each year. The paper bags were distributed through the town's public works department. Like their neighbors in Stoneham, Burlington residents recovered about 60 percent of the leaves that normally went to the landfill in the first year of the program. Both Stoneham and Burlington officials carefully examined the factors that could influence the outcome of their collection programs. In each case, they tailored the programs to the conditions in their respective towns to recover a majority of the leaves.

Table 3-1. A comparison of yard trimmings collection containers.

Type of Container	Cost	Advantages	Disadvantages
Plastic Bags	$0.12/bag	Inexpensive and readily available.	Can be torn open, scattering materials; also true for other types of bags. Require an extra debagging step because plastic can clog the tines on the turning equipment and wear out grinding blades in other machines.
		Reduce the amount of time collection vehicles spend on routes because the yard trimmings are already separated and easily handled by workers; also true for other types of bags.	
			Plastic does not decompose and is considered undesirable in the compost.
		Materials in bags are less likely to contain unwanted materials since they are not exposed; also true for other types of bags.	As grass clippings decompose in plastic bags, they will become anaerobic and therefore malodorous. Workers and nearby residents might find these odors unacceptable when these bags are opened at the composting site.
"Biodegradable" Plastic Bags	$0.20/bag	Supposed to degrade by microbial action or in the presence of sunlight, eventually becoming part of the compost.	Degradability is uncertain. Some studies have shown that these bags can take several years to fully degrade, so bits of plastic still will be visible when the compost is finished. These contaminants can reduce the marketability of finished compost.
Paper Bags	$0.25-0.45/bag	Can offer additional holding strength over lightweight plastic bags.	Can be more expensive than plastic bags.
		If paper bags get torn or crushed early in the composting process, such as in the compactor truck, the composting process is enhanced because paper bags are degradable.	
Rigid Plastic Bins	$50-60/bin	Bins are large enough to be practical yet small enough to be handled by the collection crews and residents without undue strain. Bins range in size from small, basket-sized to 30- and 90-gallon well-marked containers.	The initial costs of the bins might represent a prohibitive expenditure for some communities, however. Fees are frequently passed on to homeowners to pay for the start-up costs.
		Bins allow for neat storage of yard trimmings while awaiting collection.	Might require extra collection time to empty bins and collect materials.
		The time that yard trimmings spend in anaerobic conditions is often minimized (depending on how long the material is in the bin) since the yard trimmings are emptied from the bin and transported unbagged. This, in turn, reduces the potential for odor problems.	

Source: Wagner, 1991.

community can choose is to require residents to separate yard trimmings into color-coded or otherwise marked bags that can be sorted easily at the processing facility.

Some communities provide bags at no cost to residents and cover the cost as part of their solid waste management budget. Others sell bags to the residents at full price or at a discount. If bags are sold to residents, incentives to purchase the bags and participate in the program must be provided to discourage individuals from mixing their yard trimmings with refuse. In areas of the country that charge for general refuse collection by the barrel and maintain a bagged yard trimmings collection program, residents might be tempted to conceal noncompostable materials in composting bags as a way to decrease their own disposal costs. To minimize this problem, transparent plastic bags can be used. This strategy is being employed by a number of communities, including Brookline, Massachusetts. These bags allow sanitation workers to easily identify the contents of the bag, as well as any undesirable objects that might be readily visible. Town ordinances prohibiting the mixing of yard trimmings with refuse also might be considered. Figure 3-2 provides an example of a town ordinance.

AN ORDINANCE.

AMENDING TITLE 7, CHAPTER 7.16 OF THE REVISED ORDINANCES OF THE CITY OF SPRINGFIELD, 1986, AS AMENDED

Be it ordained by the City Council of the City of Springfield, as follows:

Title 7, Chapter 7.16 of the Revised Ordinances of the City of Springfield, 1986, as amended, is hereby further amended by inserting the following new section 7.16.041 <u>Mandatory Yard and Leaf Waste Composting.</u>

<u>7.16.041 Mandatory Leaf and Yard Waste Composting</u>

A. There is hereby established a program for the mandatory separation of certain compostable leaf and yard waste material from garbage or rubbish by the residents of the City of Springfield and the collection of these compostable leaf and yard waste materials at the residents' curbside. The collection of separated compostable leaf and yard waste material shall be made periodically under the supervision of the Director of Public Works.

B. For the purposes of this ordinance the following definitions apply:

1. Leaves - Deciduous and coniferous seasonal deposition.

2. Yard Waste - grass clippings, weeds, hedge clippings, garden waste, and twigs and brush not longer than two (2) feet in length and one-half (1/2) inch in diameter.

3. Paper Leaf Bag - A paper leaf bag shall be a Sanitary Kraft Paper Sack or equal of thirty (30) gallon capacity, two (2) ply fifty (50) pound wet strength with decomposing glue and reinforced self-supporting square bottom closure.

4. Leaf and Yard Waste Collection season - the autumn leaf season beginning the first full week of October and ending the second full week of December.

C. Separation of Compostable Leaf and Yard Waste Material and Placement for Removal.

During the Leaf and Yard Waste Collection Season Residents shall place their leaf and yard waste material into paper leaf bags as defined in Section 7.16.041.B. of barrels. These paper bags or barrels shall be place on the curbside or treebelt in accordance with section 7.16.060 on the special leaf and yard waste collection days specified by the Department of Public Works and advertised in the Springfield daily newspapers.

Figure 3-2. Mandatory yard trimmings and leaf composting ordinance from the City of Springfield, New York.

No material other than that specified in Section 7.16.041.B shall be placed in these paper bags or barrels.

Compostable leaf and yard waste material shall not be placed in plastic trash bags during the Leaf and Yard Waste Collection Season. Leaves and yard waste shall not be placed in the same refuse container as or otherwise mixed with other forms of solid waste for collection, removal, or disposal during Leaf and Yard Waste Collection Season. Any violation of this Section C or any part thereof shall be punishable by a fine not to exceed fifty dollars.

When the Owner has failed to comply with the requirements of Section C of this Ordinance, the Director of the Department of Public Works in his discretion, may refuse to collect the leaf and yard waste material and all garbage, or paper, ashes, or rubbish of the owner until the next regular pick-up, and the owner shall remove from the curb such garbage, leaf and yard waste material, and all other paper, ashes, and rubbish.

D. Ownership of Compostable Leaf and Yard Waste Materials.

Upon placement of compostable leaf and yard waste material for collection by the City at the curbside or treebelt in accordance with the special collection day, pursuant to this ordinance, such materials shall become the property of the City. It shall be a violation of this ordinance for any person, other than authorized agents of the City acting in the course of their employment, to collect or pick up or cause to be collected or picked up any compostable leaf and yard waste material so placed. Each and every such collection or pick up in violation hereof from one or more locations shall constitute a separate and distinct offense. The compostable leaf and yard waste material collected by the City shall be transported to and composted at a designated Leaf and Yard Wast Composting Site. Any violation of this paragraph D or any part thereof shall be punishable by a fine not to exceed one hundred ($100.00) dollars.

E. All ordinance, resolutions, regulations or other documents inconsistent with the provisions of this ordinance are hereby repealed to the extent of such inconsistency.

F. This ordinance and the various parts, sentences, and clauses thereof are hereby declared to be severable. If any part, sentence, or clause is adjusted invalid, it is hereby provided that the remainder of this ordinance shall not be affected thereby.

G. This ordinance shall take effect for the Leaf and Yard Waste Collection Season commencing in 1988.

Approved: October 3, 1988

Effective: October 7, 1988

Attest: *Welle Zuetzger* City Clerk

Source: Richard et al., 1990.

Figure 3-2. (Continued).

Whichever curbside collection system is used, if the containerized yard trimmings are collected on the same day as discards, provisions must be made for keeping the compostable materials separate after pickup. Compartmentalized vehicles can be used to accommodate this need; they are especially efficient if all factions of the collected material will be processed at the same facility. Since the late 1980s, a number of compartmentalized trucks have come on the market, some of which have compaction devices for each compartment (see Appendix B). Using compartmentalized trucks can avert the expense of an extra pickup crew. The amount of yard trimmings must be estimated fairly accurately, however, to prevent one compartment of the truck from filling up before the other, forcing the crew to deliver the materials before the entire vehicle is full. (Although, on average, yard trimmings constitute 18 percent of the nation's municipal discards, local factors such as climate and demographics can affect the amount of leaves or grass generated. Collection officials often have information pertaining to waste stream composition.) Another alternative that the community can choose is to require residents to separate yard trimmings into color-coded or otherwise marked bags that can be sorted easily at the processing facility.

Factors in MSW Collection

Communities that decide to collect MSW for composting can opt to source separate or commingle this material. Source-separated MSW involves varying degrees of materials segregation, which is performed where the MSW is generated. Commingled MSW is not separated by the generator. The decision to collect source-separated or commingled MSW is a significant one and affects how the material is handled at the composting facility, the preprocessing and processing costs, and the quality and marketability of the finished compost. Table 3-2 summarizes the major advantages and disadvantages of each collection method.

Source-Separated MSW

Source separation of MSW entails the segregation of compostables, noncompostables, and recyclables by individuals at the point of generation. The community then collects and transports the separated materials accordingly. Source-separation strategies can remove:

- Compostable materials, such as certain grades of paper, that can be more economically recycled than composted. In some areas, markets for certain grades of paper are strong. Therefore, a community could opt to sell collected paper for its resource value rather than compost it.

- Noncompostable recyclables such as aluminum, glass, and plastic beverage containers.

> ### Avoiding Undesirable Materials in Feedstock Collections
>
> Both yard trimmings and collected MSW can contain materials that might affect processing and product quality. These materials can include glass, metals, beverage containers, plastics, household hazardous waste, and other undesirable materials. Collecting crews should be trained to recognize and separate these types of materials whenever possible. Because of the variety of materials collected, MSW feedstock is likely to contain larger amounts of undesirable materials than yard trimmings feedstock. Although yard trimmings can contain pesticides and herbicides commonly used by residents and businesses, the composting process will break down many of these substances, limiting their impact on the final product (see Chapter 6 for a more detailed discussion).
>
> Communities can take steps to reduce the amount of undesirable materials in the feedstock. These include passing ordinances, posting warning notices, and issuing fines for mixing noncompostables with compostables. In addition, bagged yard trimmings and MSW bins can be opened at the curb to detect undesirable materials. Facility employees can look for and separate out unwanted materials (see Chapter 4).

- Materials that are difficult to compost such as brush.

- Household hazardous waste such as paints, batteries, pesticides, and used oil.

- Noncompostable nonrecyclables such as light bulbs and toothpaste tubes.

The primary benefit of source separation is that the feedstock tends to contain fewer unwanted materials, particularly heavy metals (Glaub et al., 1989). In addition, source separation can help remove those items from the waste stream that are difficult to separate at the facility, such as plastic, which is often shredded; and glass, which can shatter into small, hard-to-remove pieces. This produces a higher quality compost. Most MSW composting facilities in communities with source-separation programs perform an additional sorting of incoming materials to produce a still cleaner compost feedstock. Communities with MSW composting facilities can combine source separation of compostable materials with source separation of other recyclable materials such as glass, aluminum, and plastic.

A study conducted in 1990 revealed that a majority of MSW composting facilities prefer processing source-separated over commingled MSW (Goldstein and Spencer, 1990). The study indicated that recycled materials are cleaner and more marketable if source separated since they

Table 3-2. Source separation vs. commingling of MSW.

Advantages	Disadvantages
Source Separation of MSW	
Less chance of collecting unwanted objects, which can result in a higher quality compost product.	Can be less convenient to residents.
Less money and time spent on handling and separation at facility.	Might require the purchase of new equipment and/or containers.
Provides an educational benefit to residents and might encourage source reduction.	Might require additional labor for collection.
Collection of Commingled MSW	
Usually done with existing equipment and labor resources.	Higher potential for collecting unwanted objects, which can result in a lower quality compost product.
Convenient to residents since no separation is required.	Higher processing and facility costs.

are not mixed with undesirable materials. Moreover, the amount of noncompostable material received at the composting facility is reduced. This means fewer noncompostables must be separated out on site and sent to landfills or recycling centers, resulting in lower transportation and labor expenditures. Finally, the quality and appearance of the compost can be improved and therefore command a higher price. (Chapter 4 discusses the role of source separation on preprocessing at the composting facility in more detail; Chapter 9 discusses the role that source separation can play in reducing heavy metals and other contaminants in the final compost product.)

Source separation of MSW for composting can be done in bins or bags. Some programs require that compostables, noncompostables, and recyclables be placed in different bins for curbside collection. While a large number of collection containers can be unsightly to some citizens, the containers themselves are usually small since each one holds small volumes of materials. Some municipalities even use small baskets (similar to milk crates) to collect glass, paper, and metals.

While source separation can avert many of the expenses associated with preprocessing compostables, other costs must be considered. The community very likely will have to devote more labor to the collection process. In addition, containers or bins must be purchased either by the municipality or citizens. The degree of participation is a variable also, so a thorough public education and awareness campaign is necessary to encourage residents and businesses to separate out noncompostables (see Chapter 10).

Commingled MSW

Commingled MSW collection is the method that municipalities traditionally have used to pick up materials from residents and businesses. Commingling allows residents to combine trash, compostables, and recyclables in the same containers. The municipality then collects and transports the materials to the composting facility. Commingled MSW collections usually can be done with existing equipment. Collection time and cost per ton often are less than

Wet/Dry Separation Strategies for Composting

Some communities in Canada and Europe are using or experimenting with the separation of materials into wet and dry components. The City of Guelph in Ontario, Canada, reported a diversion rate of more than 60 percent using this collection strategy (Hoornweg et al., 1991).

The wet stream includes all organic kitchen scraps, yard trimmings, nonrecyclable paper, and some noncompostable elements. The dry stream comprises all dry noncompostables and recyclables. The dry waste stream is sent to a landfill or materials recovery facility (MRF) where recyclables are removed for recovery. Wet materials are sent to a compost facility.

Since 1989, Guelph has been conducting a pilot program to test four different materials separation techniques in over 500 households. The city has found that the highest diversion rates were achieved by citizens dividing the MSW stream into wet and dry components and placing these components in green and blue plastic bins, respectively. The city currently is investigating other aspects of the program, including separation in multi-family dwellings and commercial and educational institutions.

As of August 1993, plans were underway to open a 139,000 ton per year facility, including a 44,000 ton per year "wet" composting plant and an 85,000 ton per year "dry" MRF (Darcey et al., 1993).

those for separated materials since sanitation collectors can fit more into single unit packer trucks at a faster rate. Commingled commercial materials are deposited in large metal or plastic bins equipped with hinged lids. These bins are designed for easy transport to the processing facility. Some bins are equipped with a compactor, making it possible to increase the capacity of each container. Compaction can make separation more difficult, however, and can greatly complicate the procedures and equipment that will be used to compost.

The primary disadvantage of a commingled MSW collection program is that the separation must be performed as soon as possible once the material arrives at the facility. At the facility, the organic materials are typically separated by both manual and mechanical means (see Chapter 4) in order to remove them from the recyclable and other noncompostable materials—a process that requires significant labor and specialized equipment. Additionally, commingling does not require individuals to change their behavior thereby becoming more aware of the resource value of materials they discard.

Summary

Whether designing a yard trimmings or MSW composting program, collection is a key factor in ensuring the program's success. Not only does collection have a direct bearing on the willingness of households to participate in and endorse a program but the collection strategy chosen also influences the way that the feedstock is handled and processed at the facility as well as the quality and marketability of the final product. Additionally, collection can be one of the most expensive aspects of a composting program and influences labor, equipment, processing, and other resource needs. For these reasons, decision-makers should carefully examine and weigh all possible collection methods to determine the best approach for their community.

Chapter Three Resources

Appelhof, M., and J. McNelly. 1988. Yard waste composting guide. Lansing, MI: Michigan Department of Natural Resources.

Ballister-Howells, P. 1992. Getting it out of the bag. BioCycle. March, 33(3):50-54.

Cal Recovery Systems (CRS) and M. M. Dillon Limited. 1989. Composting: A literature study. Ontario, Canada: Queen's Printer for Ontario.

Darcey, S. 1993. Communities put wet-dry separation to the test. World Wastes. 36(98):52-57.

Glaub, J., L. Diaz, and G. Savage. 1989. Preparing MSW for composting. As cited in: The BioCycle Guide to Composting Municipal Wastes. Emmaus, PA: The JG Press, Inc.

Glenn, J. 1992. Integrated collection of recyclables and trash. BioCycle. January, 33(1):30-33.

Glenn, J. 1989. Taking a bite out of yard waste. BioCycle. September, 30(9):31-35.

Goldstein, N., and B. Spencer. 1990. Solid waste composting facilities. BioCycle. January, 31(1):36-39.

Hoornweg, D., L. Otten, and W. Wong. 1991. Wet and dry household waste collection. BioCycle. June, 32(6): 52-54.

Mielke, G., A. Bonini, D. Havenar, and M. McCann. 1989. Management strategies for landscape waste. Springfield, IL: Illinois Department of Energy and Natural Resources, Office of Solid Waste and Renewable Resources.

Richard, T., N. Dickson, and S. Rowland. 1990. Yard waste management: A planning guide for New York State. Albany, NY: New York State Energy, Research and Development Authority, Cornell Cooperative Extension, and New York State Department of Environmental Conservation.

U.S. Environmental Protection Agency (EPA). 1989. Decision-Maker's Guide to Solid Waste Management. EPA/530-SW-89-072. Washington, DC: Office of Solid Waste and Emergency Response.

Wagner, T.C. 1991. In search of the perfect curbside system. BioCycle. August, 32(8).

Wirth, R. 1989. Introduction to composting. St. Paul, MN: Minnesota Pollution Control Agency, Ground Water and Solid Waste Division.

Chapter Four
Processing Methods, Technologies, and Odor Control

This chapter describes the three stages of composting (preprocessing, processing, and postprocessing) for both yard trimmings and MSW composting. It examines the operations that must be performed at each step in the process and describes ways for optimizing those conditions that influence the process. In addition, this chapter discusses the different technologies currently used to compost yard trimmings and MSW feedstocks in the United States. These can range from simple, low-technology systems that require minimal attention and maintenance to complex systems that use sophisticated machinery and require daily monitoring and adjustment. The design and complexity of a composting operation are determined by the volume, composition, and size distribution of the feedstock; the availability of equipment; the capital and operating funds; and the end-use specifications for the finished product. This chapter also examines the potential problems associated with odor and describes the measures a composting facility can take to prevent or minimize odor. A system flow chart for a typical operation that composts yard trimmings is shown in Figure 4-1. Figure 4-2 outlines a process flow diagram for a typical MSW composting facility. For more information on costs and effectiveness of the equipment described in this chapter, see Appendix B. Two case studies illustrating the process of composting yard trimmings and MSW are included at the back of this chapter.

Preprocessing

During preprocessing, feedstock is prepared for composting. Preprocessing has a significant impact on the quality of the finished compost product and the speed at which processing can be conducted. In general, the more effective the preprocessing, the higher the quality of the compost and the greater the efficiency of processing. Three procedures are typically performed during preprocessing: 1) sorting feedstock material and removing materials that are difficult or impossible to compost; 2) reducing the particle size of the feedstock material; and 3) treating feedstock to optimize composting conditions. These composting procedures are described below for both yard trimmings and MSW.

Sorting

The level of effort required to sort and remove unwanted materials from the composting feedstock depends on several factors, including the source of the feedstock, the end use of the product, and the operations and technology involved. The more diverse the feedstock material, the more sorting and removal will be required. For this reason, yard trimmings (which tend to be relatively uniform) generally require little sorting while MSW (which comprises heterogeneous materials) can require extensive sorting and separation. The end-use specifications for the finished compost product also affect the level of effort involved as some end uses require a higher quality product than others. For example, compost that will be used as landfill cover can have higher levels of unwanted materials than compost that will be used on food crops. Composting operations designed to produce landfill cover can therefore utilize simpler and less thorough sorting and removal methods.

Sorting Techniques for Yard Trimmings Feedstock

Upon delivery to a composting site, yard trimmings should be visually inspected to detect any materials that could affect the composting process. Visual inspection can be readily accomplished by spreading out the material on

Processing Methods, Technologies, and Odor Control

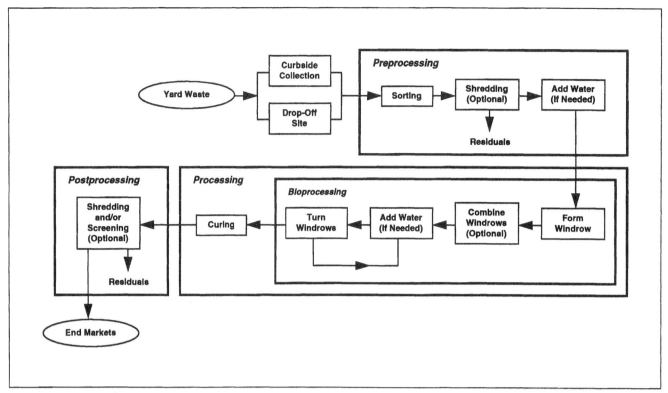

Figure 4-1. Typical yard trimmings composting operation.

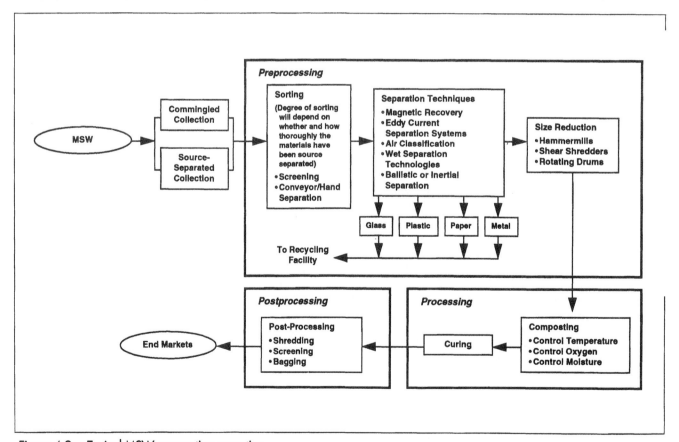

Figure 4-2. Typical MSW composting operation.

the tipping floor where the feedstock is unloaded. Workers can then physically remove any undesirable objects present. Materials that should be removed are those that would interfere with mechanical composting operations, inhibit the decomposition process, cause safety problems for those working with or using the compost, or detract from the overall aesthetic value of the finished compost product. Plastic bags are the chief problem at most yard trimmings composting facilities.

Feedstock with a significant amount of unwanted objects can be hand-sorted more efficiently with a mechanical conveyor belt. With this approach, the feedstock material is loaded into a hopper that discharges at a slow speed onto a conveyor belt. Workers on either side of the moving belt manually pick out glass, plastic, and other visible noncompostables. To facilitate sorting, the belt width should allow the workers to reach the center of the belt, and the trimmings should not be more than 6 inches deep. Materials removed from the conveyor belt are deposited into storage containers that can be moved easily to other storage/processing areas. These noncompostable materials are considered residuals from the composting process and generally are recycled or disposed of by landfilling.

For reasons of health and safety, it is important that workers avoid physical contact with undesirable materials during manual sorting and removal. The sorting area should be well-lit and properly ventilated, and the conveyor belt should be set up to minimize motion injuries such as back strain. Those handling the materials should wear heavy gloves and follow specified hygiene practices (see Chapter 6 for more information on worker health and safety).

Sorting Techniques for MSW Feedstock

In general, sorting of MSW prior to composting requires more labor and machinery than sorting yard trimmings because of the diversity of MSW. As mentioned earlier, MSW is extremely heterogeneous in size, moisture, and nutrient content, and the organic fractions can contain varying degrees of noncompostable and possibly hazardous waste. Both physical and chemical materials found in the feedstock can have a negative impact on the marketability of the finished product, and their removal forms a large part of the expense of modern MSW composting facilities (Richard, 1992). Both manual and mechanical techniques can be used to sort feedstock materials and remove unwanted items.

Many items in the MSW composting feedstock are recyclable, such as aluminum cans, ferrous materials, and plastic bottles. Because of the potential value of these recyclables, the separation, removal, and collection of these items should be pursued. Although the MSW feedstock can be sorted after being subjected to size-reduction processes, it is advisable to remove recyclables before size reduction (this also will improve the value of recyclables).

Sorting before size reduction also will prevent recyclables from being pulverized and mixed into the feedstock, which can cause a variety of problems. For example, plastics are difficult to remove after they are shredded and mixed with compostable materials. Shattered glass generates shards that can remain in the compost and devalue the finished product as well as present a safety hazard both to workers sorting the compost and to compost users.

Materials targeted during manual separation include recyclables and inert materials. As in the case of yard trimmings, manual separation along a conveyor belt represents the most effective method to remove noncompostable materials and chemicals from feedstock. Health and safety provisions for manually sorting are particularly important in the case of MSW feedstock, which might contain potentially dangerous items such as syringe needles, pathogenic organisms, broken glass, or other materials that could cause injury or infection (see Chapter 6).

Mechanical sorting and removal techniques are based on the magnetic and physical (i.e., weight and size) properties of the feedstock materials. Magnetic-based systems separate ferrous metals from the rest of the feedstock; eddy-current machines separate out nonferrous metals; size-based systems such as screens separate different sizes of materials; and weight-based systems separate out heavier noncompostable materials such as metals, glass, and ceramics.

Table 4-1 outlines mechanical separation technologies that are currently used in MSW composting. These technologies are discussed briefly below and in more detail in Appendix B.

- *Screens* - Screens are used in most MSW composting facilities to control the maximum size of feedstock and to separate materials into size categories. The main purpose of this size fractionation is to facilitate further separation. Screens separate small dense materials such as food scraps, glass, and small, hard plastic pieces from the bulky, light fraction of the feedstock. The type of screen used depends on the moisture content, cohesiveness, heterogeneity, particle shape, and density of the feedstock to be segregated. Trommel screens are commonly used for initial materials processing at MSW facilities. Figure 4-3 illustrates a trommel screen.

- *Magnetic-based separators* - Magnetic separators create magnetic fields that attract ferrous metals and remove them from the rest of the feedstock stream as it travels along conveyors. Magnetic separators are among the most effective and inexpensive unit processes available for sorting and removing contaminants from the feedstock. The economic benefits of these devices are enhanced by selling the

Table 4-1. Processing MSW feedstock—separation techniques.

Technology	Materials Targeted
Screening	Large: Film plastics, large paper, cardboard. Mid-sized: Recyclables, most organics. Fine: Organics, metal fragments.
Magnetic Separation	Ferrous metals.
Eddy-Current Separation	Nonferrous metals.
Air Classification	Light: Paper, plastic. Heavy: Metals, glass, organics.
Wet Separation	Floats: Organics. Sinks: Metals, glass, gravel.
Ballistic Separation	Light: Plastic, undecomposed paper. Heavy: Metals, glass, gravel.

Source: Richard, 1992.

scrap metals these units separate from the compostable materials. The efficiency of magnetic separators depends primarily on the quantity of materials processed and the speed at which they pass through the magnetic field. The size and shape of the ferrous objects, as well as the distance between the magnet and the objects, also are important variables. To increase the efficiency of the separation process, more than one magnetic separation technology can be used in series with another. Applying air classification (described below) prior to magnetic separation minimizes the contaminants in the scrap ferrous even further.

- *Eddy-current machines* - Eddy-current machines separate aluminum and other nonferrous metals from MSW. These machines generate a high-energy electromagnetic field that induces an electrical charge in nonferrous metals (and other materials that conduct electricity). The electrical charge forces these materials to be repelled from non-charged fractions of the feedstock material. The feedstock should be conveyed to eddy-current machines after magnetic separation to minimize contamination by ferrous metals. Recovery rates for eddy-current separators vary with the depth of the material on the conveyor belt, belt speed, the degree of preprocessing, and the strength of the magnetic field. Full-scale trials and manufacturer estimates of separation efficiency in MSW applications range from 50 to 90 percent. Figure 4-4 illustrates an eddy current separator.

- *Air classifiers* - Air classifiers separate feedstock materials based on weight differences; for example, the heavier fractions (metals, glass, ceramics, and rocks) are removed from the lighter materials. The heart of an air classification system is an air column or throat into which the materials stream is fed at a metered rate. A large blower sucks air up through the throat, carrying light materials such as paper and plastic. These then enter a cyclone separator where they lose velocity and drop out of the air stream. Heavy materials fall directly out of the throat. An important consideration when using air classifiers is that although most of the heavier materials separated out are noncompostable, some materials that fall out (e.g., certain food materials and wet paper) can be composted (Glaub et al., 1989). Air classifiers typically are used after the feedstock has been size-reduced. Separation efficiency in experimental application of air classification systems has reached 90 percent for plastics and 100 percent for paper materials. In combination with screening and size reduction, air classification can be used to significantly reduce metal contaminant levels. Figure 4-5 illustrates an air classification system.

- *Wet separation technologies* - Wet separation technologies are similar to air classification systems in that they separate materials based upon density, but water replaces air as the floating medium in these technologies. After entrainment in a circulating water stream, the heavy fraction drops into a sloped tank where it moves to a removal zone. The lighter organic matter floats and is removed from the recirculating water using stationary or rotating screening systems similar to those employed by wastewater treatment facilities. This technology is particularly effective for removing glass fragments and other sharp objects.

- *Ballistic or inertial separation* - This technology separates inert and organic constituents based upon density and elasticity differences. Compost feedstock is dropped on a rotating drum or spinning cone and the resulting trajectories of glass, metal, and stones, which depend on density and elasticity, bounce the materials away from the compost feedstock at different lengths. Figure 4-6 illustrates a ballistic separator.

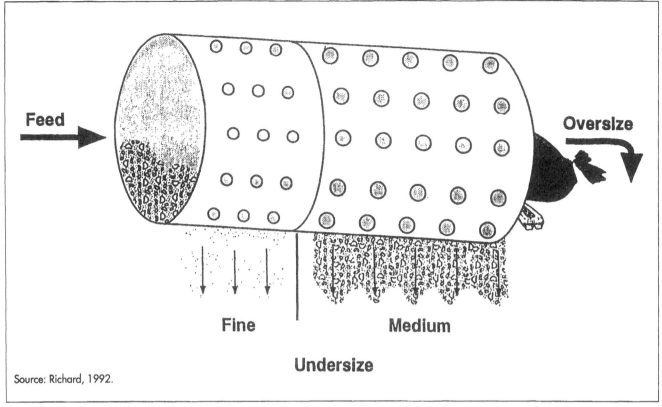

Figure 4-3. Trommel screen.

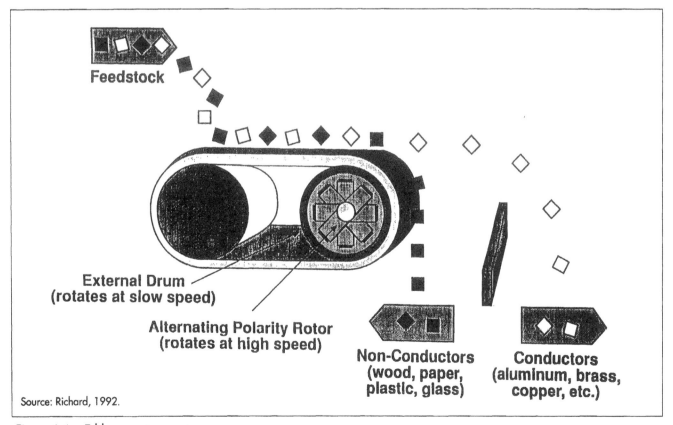

Figure 4-4. Eddy-current separator.

Processing Methods, Technologies, and Odor Control

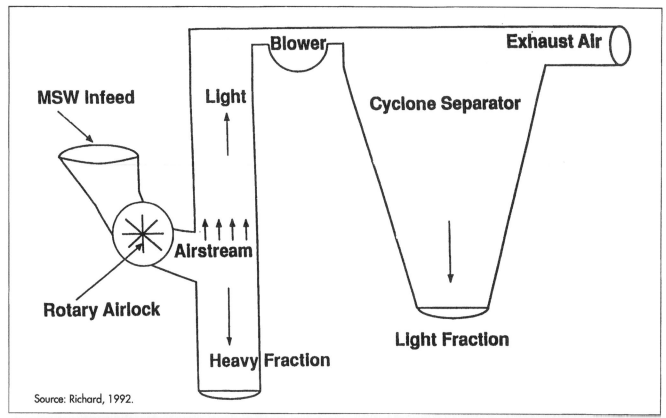

Figure 4-5. Air classification system.

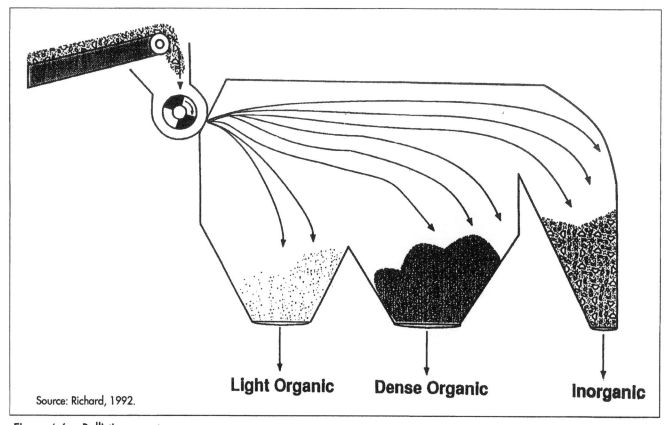

Figure 4-6. Ballistic separator.

Reducing the Particle Size of the Feedstock

Size reduction usually is performed after noncompostables have been separated from the compostable feedstock. Some separation technologies, including magnetic separation, air classification, and wet separation, achieve greater levels of removal only after size reduction, however. The exact order of steps varies in different composting operations depending on the type and volume of feedstock to be composted. Proper sequencing of these preparation processes can have a significant impact on system performance.

The primary reason for performing size reduction is to increase the surface area to volume ratio of the feedstock materials. This enhances decomposition by increasing the area in which microorganisms can act upon the composting materials. If composting materials are too small, however, air flow through the compost pile will be reduced. This reduced oxygen availability has a negative impact on decomposition. Maximizing composting efficiency requires establishing a balance between reducing particle size and maintaining aerobic conditions. A study of the tradeoff between increased surface area for decomposition and reduced pore size for aeration concluded that particle sizes of 1.3 to 7.6 cm (0.5 to 3.0 inches) are most efficient (Gray and Biddlestone, 1974). The lower range is suitable for forced aeration systems while the larger range is preferred for windrows and other systems supplied with oxygen by passive diffusion and natural convection.

Yard Trimmings

Size reduction of most types of yard trimmings can help accelerate the composting process. Size reduction is warranted for woody material mixed with other yard trimmings since wood decomposes at a very slow rate and might delay the development of the compost end product. Some facilities have found that shredding leaves as well will reduce the time required to produce finished compost from 18 months to 9 months (Richard et al., 1990). Excessive size reduction of leaves and grass could prove undesirable, however, because small particles can inhibit aerobic conditions and impede release of heat from the composting masses. If grass clippings become compacted, they can restrict oxygen flow and create anaerobic pockets in the composting mass. Finely shredded yard trimmings must be turned more frequently to prevent these anaerobic conditions. Tub grinders are a common piece of size reduction machinery at large facilities for composting yard trimmings. These grinders use a rotating tub to feed a horizontal hammermill (see following section).

MSW

Size reduction homogenizes MSW feedstock materials, achieving greater uniformity of moisture and nutrients to encourage even decomposition. A variety of size-reduction devices are available, the most common of which are hammermills, shear shredders, and rotating drums. This equipment is outlined below and described in more detail in Appendix B.

- *Hammermills* - Hammermills reduce the size of feedstock materials by the action of counter rotating sets of swinging hammers that pound the feedstock into smaller sized particles. The hammer axles can be mounted on either a horizontal or a vertical axis and usually require material to pass through a grate before exiting. Mills that lack the exit grate are termed flail mills. Figure 4-7 illustrates a hammermill.

- *Shear shredders* - Shear shredders usually consist of a pair of counter rotating knives or hooks that rotate at a slow speed with high torque. The shearing action tears or cuts most materials, which helps open up the internal structure of the particles and enhances opportunities for decomposition.

- *Rotating drums* - Rotating drums use gravity to tumble materials in a rotating cylinder. Material is lifted by shelf-like strips of metal along the sides of the drum, which can be set on an incline from the horizontal. Some of the variables in drum design include residence time (based on length, diameter, and material depth), inclination of the axis of rotation, and the shape and number of internal vanes (which lift materials off of the bottom so they can fall through the air). Figure 4-8 illustrates a rotating drum.

If materials such as gas cylinders and ignitable liquids are present in MSW feedstock, there is a potential for explosions during size reduction. Visual inspection, along with sorting and removal procedures, can minimize this potential. Nevertheless, size reduction equipment should be isolated in an explosion-proof area within the composting facility, and proper ventilation for pressure relief should be provided.

Treating Feedstock Materials to Optimize Composting Conditions

To enhance composting, both yard trimmings and MSW feedstock can be treated before processing. Such treatment can optimize moisture content, carbon-to-nitrogen (C:N) ratio, and acidity/alkalinity (pH). (These parameters were introduced in Chapter 2.)

Moisture Content

Maintaining a moisture content within a 40 to 60 percent range can significantly enhance the composting process. Before composting begins, the feedstock should be tested for moisture content. The "squeeze test" is a simple method of determining whether the moisture content falls within the proper range. If just a few drops of water are released from a handful of the feedstock when squeezed, the moisture content is acceptable. If a more definitive determination of moisture content is needed, a sample of the feedstock can be weighed, oven-dried at about 104°C (219°F) for 8 hours, and weighed again. The moisture content can be derived by the following formula:

moisture content = (wet weight − dry weight)/wet weight

With yard trimmings, the moisture content of leaves tends to be lower than optimal. The moisture content of grass tends to be higher than optimal. Moisture, therefore, should be added to dry leaves, generally at a level of about 20 gallons of water per cubic yard of leaves (Richard et al., 1990). During the early stages of composting, leaves must be mixed during wetting to prevent the water from running off the pile surface. On the other hand, grass should be mixed with drier materials (such as leaves or wood chips) or turned more frequently during the initial stages of processing to facilitate the evaporation of excess water.

Moisture content in the MSW feedstock varies widely. Significant attention, therefore, should be paid to assessing moisture levels of MSW and mixing materials streams to optimize moisture content of the composting feedstock. For high-rate MSW composting, a minimum moisture content of 50 to 55 percent is recommended (Golueke, 1977). Since MSW feedstock is often drier than this, water must be added during the composting and curing stages to bring the moisture content into the optimal range. MSW compost mixtures usually start at about 55 percent moisture and dry to 35 percent moisture (or less) prior to final screening and marketing (CC, 1991).

Mechanical aeration and agitation directly influence the moisture content of the composting pile. Aeration increases flow through the composting pile, inducing evaporation from the interior spaces. Turning composting piles exposes the interior of the piles, releasing heated water as steam. This moisture loss can be beneficial, but if excess moisture is lost (i.e., the moisture content falls to 20 percent), rewetting might be required (Richard, 1992). MSW composting piles usually require additional water.

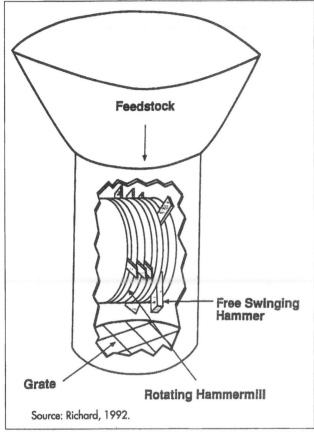

Figure 4-7. Hammermill.

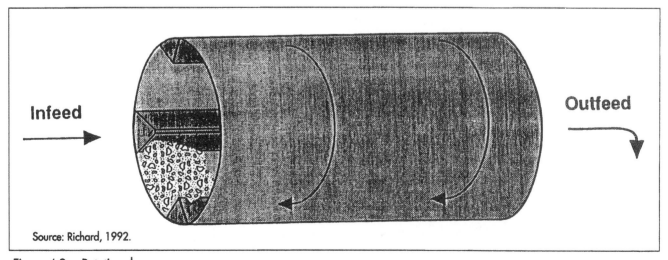

Figure 4-8. Rotating drum.

Finally, temperature determines how much moisture will be lost with turning and aeration; the higher the temperature, the more water will be lost via evaporation. In turn, moisture loss affects the temperature of the piles.

Carbon-to-Nitrogen (C:N) Ratio

Most of the nutrients needed to sustain microbial decomposition are readily available in yard trimmings and MSW feedstocks. However, carbon and nitrogen might not be present in proportions that allow them to be used efficiently by microorganisms. Composting proceeds most efficiently when the C:N ratio of the composting material is from 25:1 to 35:1. When the C:N ratio is greater than 35:1, the composting process slows down. When the ratio is less than 25:1, there can be odor problems due to anaerobic conditions, release of ammonia, and accelerated decomposition.

Generally, the C:N ratio for yard trimmings can be approximated by examining the nature of the feedstock; green vegetation is high in nitrogen and brown vegetation is high in carbon. While the diversity of MSW feedstock material makes an estimation of the C:N ratio somewhat difficult, a precise C:N ratio can be determined by laboratory analysis. Feedstock materials with different C:N ratios can be mixed to obtain optimal levels of carbon and nitrogen when necessary. (see Table 4-2 for carbon-to-nitrogen ratios for various organic materials).

Acidity/Alkalinity (pH)

The closer the pH of the feedstock material is to the neutral value of 7, the more efficient the composting process will be. Fresh leaves tend to have pH levels of approximately 7 (Strom and Finstein, 1989). Fruit scraps generally are acidic with a pH below 7 (CRS, 1989). Kits to test pH levels are readily available and easy to use. If pH levels are significantly higher than 8 (an unusual situation), acidic materials, such as lemon juice, can be added to the feedstock. If the feedstock has a pH significantly below 6, buffering agents, such as lime, can be added. Because pH levels are largely self-regulating, actions to bring pH to optimum levels are rarely necessary (CRS, 1989; Strom and Finstein, 1989).

Mixing

Mixing is often required to achieve optimal composting conditions. Mixing entails either blending certain ingredients with feedstock materials or combining different types of feedstock materials together. For example, bulking agents (such as wood chips) are often added to feedstock materials that have a fine particle size (such as grass). Bulking agents have the structural integrity to maintain adequate porosity and help to maintain aerobic conditions in the compost pile. Bulking agents are dry materials and tend to have a high carbon content. Therefore,

Table 4-2. Carbon-to-nitrogen ratio of various materials.

Type of Feedstock	Ratio
High Carbon Content	
Bark	100-130:1
Corn Stalks	60:1
Foliage	40-80:1
Leaves and Weeds (dry)	90:1
Mixed MSW	50-60:1
Paper	170:1
Sawdust	500:1
Straw (dry)	100:1
Wood	700:1
High Nitrogen Content	
Cow Manure	18:1
Food Scraps	15:1
Fruit Scraps	35:1
Grass Clippings	12-20:1
Hay (dry)	40:1
Horse Manure	25:1
Humus	10:1
Leaves (fresh)	30-40:1
Mixed Grasses	19:1
Nonlegume Vegetable Scraps	11-12:1
Poultry Manure	15:1
Biosolids	11:1
Weeds (fresh)	25:1
Seaweed	19:1

Sources: Golueke, 1977; Richard et al., 1990; Gray et al., 1971b.

whenever bulking agents are used, care should be taken to ensure that C:N ratios do not become too high.

Mixing is most efficient when it is conducted after feedstock sorting and size reduction and before processing begins. This can minimize the quantity of materials that must be mixed because noncompostables have been removed. In addition, once piles have been formed for processing, adequate mixing becomes extremely difficult.

For simple composting operations that do not require high levels of precision, mixing can be performed during size reduction or pile formation by feeding different ingredients or types of materials into these operations. When higher levels of precision are required, mixing equipment (such as barrel, pugmill, drum, and auger mixers) can be used (see Appendix B). Most mixers also compress materials, which can reduce pore space in the feedstock and inhibit aeration in the compost pile. Mixers also have relatively high capital and operating and maintenance costs so it might be impractical for smaller facilities to use them, particularly those that compost only yard trimmings.

Processing

After yard trimmings and MSW feedstock materials are preprocessed, they can be introduced into the compost processing operations. During processing, various methods can be employed to decompose the feedstock materials and transform them into a finished compost product. Processing methods should be chosen to maximize the speed of the composting process and to minimize any negative effects, such as odor release and leachate runoff.

The level of effort required for processing composting feedstock depends on the nature of the feedstock, the desired speed of production, the requirements for odor and leachate control, and the quality requirements for the finished compost. A facility's financial resources and available space also are important. In general, the greater the speed of the process, the more odor and leachate control necessary. Where greater space or level of effort is needed, more financial resources will be required.

In general, more resources and higher levels of effort are necessary to compost a MSW feedstock than a yard trimmings feedstock, largely because of the diverse nature of MSW. For composting either yard trimmings or MSW, processing occurs in two major phases: the composting phase and the curing phase. These stages are discussed below.

The Composting Stage

Microorganisms decompose the readily available nutrients present in the feedstock during composting. Because most of the actual change in the feedstock occurs during this stage, the most intensive methods and operations tend to be used here. Compost processing can occur in simple environments that are completely subject to external forces or in complex and highly controlled environments. The composting methods currently employed are (in order of increasing complexity):

- Passive piles
- Turned windrows
- Aerated static piles
- In-vessel systems

Passive Piles

Although this method is simple and generally effective, it is not applicable under all conditions or to all types of materials. Composting under these conditions is very slow and is best suited to materials that are relatively uniform in particle size. Although passive piles theoretically can be used for composting either yard trimmings or MSW, the propensity for odor problems renders them unsuitable for MSW feedstock materials or even large quantities of grass or other green materials that have a high nitrogen content.

Passive piles require relatively low inputs of labor and technology. They consist of piles of composting material that are tended relatively infrequently, usually only once each year. Tending the piles entails turning them (i.e., physically tearing down and reconstructing them). Figure 4-9 illustrates the proper method of turning a compost pile. Such an effort requires only a few days' use of personnel and equipment, making this a relatively low-cost composting method.

Before piles are turned, the moisture content of internal and external layers of the compost pile should be checked using the methods discussed in the preprocessing section of this chapter. If the moisture content is too low, water can be added by manually spraying the pile with hoses or by using automatic sprinklers or irrigation systems. If the moisture content is too high, turning can be conducted more frequently to increase evaporation rates.

With all composting methods, regular monitoring of the temperatures of composting materials is recommended. A variety of long-stem (3-foot) digital and dial-type thermometers and infrared scanners are available that can read temperatures up to 93°C (199°F).

Passive piles should be constructed large enough to conserve sufficient heat but not so large as to overheat. If temperatures of the composting mass exceed 60°C (140°F), composting materials can combust, and/or microorganisms needed for decomposition can be killed. Compost piles should be turned if this temperature is exceeded.

Even if temperature and moisture are not monitored with the passive pile composting method, the periodic turning of the piles will adjust the oxygen level, moisture content, and temperature to some degree. The movement created by turning aerates the pile, and the anaerobic center is replaced with oxygen-rich external layers of the material. In addition, dry internal materials are exposed to the outer layers of the pile where they are more susceptible to wetting by rain or snow. The increased aeration and wetting caused by turning also serve to reduce temperatures in the internal layers, preventing excessive heat buildup. Temperature and oxygen levels also can be controlled somewhat by forming piles of the appropriate size. The larger the pile, the greater the insulation and the higher the temperature levels that can be reached. The larger the pile, however, the lower the degree of oxygen penetration and the greater the potential for anaerobic conditions forming in the center of the pile.

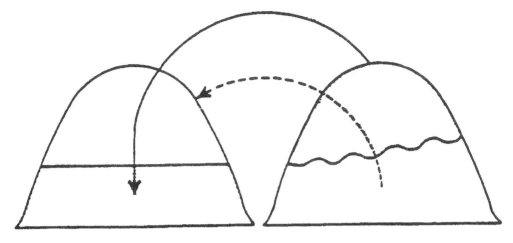

Figure 4-9. Pile turning for aeration and mixing.

Several disadvantages are associated with passive pile methods. Unlike more intensive composting processes that can produce a finished product in a few weeks to a few months, passive piles can require over 1 year for the composting process to be completed. In addition, the minimal turning of passive piles results in the formation of anaerobic conditions so that when piles are eventually turned (especially for the first year or two of the process) significant odors result. Passive piles consequently cannot be placed in densely populated areas, and a large buffer zone is recommended between residents and composting operations (Strom and Finstein, 1989). The untended passive piles also might resemble dump sites to community members who might discard trash at the site. Some means of controlling access to the passive pile site is, therefore, recommended. Finally, large, untended piles have the potential to overheat and combust, creating a possible fire hazard.

Turned Windrows

Turned windrows are a widely used method for composting yard trimmings and MSW. This method generally is not appropriate, however, for MSW containing significant amounts of putrescible materials due to odor concerns.

Turned windrows are elongated composting piles that are turned frequently to maintain aerobic composting conditions. The frequent turning promotes uniform decomposition of composting materials as cooler outer layers of the compost pile are moved to inner layers where they are exposed to higher temperatures and more intensive microbial activity. The turned windrow method results in the completion of the composting process for yard trimmings in approximately 3 months to 1 year (UConn CES, 1989).

Turned windrow operations generally can be conducted outdoors. To increase the operator's ability to control composting conditions, however, windrows can be placed under or inside shelters. Leachate problems should be minimized by constructing windrows on firm surfaces surrounded by vegetative filters or trenches to collect runoff (see Chapter 6). (A paved surface might be helpful, depending on the size and location of the facility and how muddy it might get.) Run-on controls also are helpful as is careful balancing of the C:N ratio. Progressive decomposition of the composting materials reduces the size of the windrows, allowing them to be combined to create space for new windrows or other processes.

As with passive piles, forming windrows of the appropriate size helps maintain appropriate temperature and oxygen levels. The ideal height for windrows is from 5 to 6 feet (CRS, 1989). This height allows the composting materials to be insulated properly but prevents the buildup of excessive heat. Windrow heights vary, however, based on the feedstock, the season, the region in which the composting operation is being conducted, the tendency of the composting materials to compact, and the turning equipment that is used. Windrow widths generally are twice the height of the piles. Factors such as land availability, operating convenience and expedience, type of turning equipment used, and interest in the end product quality also affect the chosen windrow width. Careful monitoring of width is unnecessary, however, to ensure that proper oxygen and temperature levels are maintained; windrow height determines aeration levels to a far greater degree than windrow width. Windrow length also has little impact on the composting process.

Windrow shapes can be altered to help maintain appropriate composting conditions (primarily moisture levels). For example, windrows with concave crests are appropriate during dry periods and when the moisture content of the composting material is low to allow precipitation to be captured more efficiently. Peaked windrows are preferable during rainy periods to promote runoff of excess water and to prevent saturation. Illustrations of these windrow shapes are presented in Figure 4-10.

The same types of operations used to monitor critical composting conditions in the passive pile method also can be used with turned windrow composting. The more frequent turning of composting materials with the turned windrow technique does tend to maintain oxygen, moisture, and temperature at appropriate levels, however. Where odor control and composting speed are a high priority, oxygen monitoring equipment can be installed to alert operators when oxygen levels fall below 10 to 15 percent, which is the oxygen concentration required to encourage aerobic decomposition and minimize odor problems (Richard, 1992).

Turning frequencies for this method can range from twice per week to once per year. In general, the more frequently that the piles are turned, the more quickly the composting process is completed. Some materials do not need to be turned as frequently to maintain high levels of decomposition. For example, structurally firm materials have greater porosity and therefore can maintain aeration for greater periods of time without turning. Ideal turning patterns should move the outside layers of the original windrow to the interior of the rebuilt windrow (this pattern is shown in Figure 4-11). If this pattern is not feasible, then care should be taken to ensure that all materials spend sufficient time in the interior of the pile. Inefficiencies in the turning pattern can be compensated for by increasing the frequency with which the windrows are turned.

The turning equipment used will, in part, determine the size, shape, and space between the windrows. Front-end loaders are commonly used in smaller operations. The quantity of materials that they can handle as well as the control that they can exercise over the turning process is limited, however. When this equipment is used, enough space must be maintained between windrows to allow the front-end loaders to maneuver and turn the piles. Windrow turners are larger machines that are often used at

Landspreading

Landspreading involves the placement of organic materials on the ground for decomposition under uncontrolled conditions. A few simple interventions, however, such as reducing feedstock particle size or periodically turning materials with a plow, can be used to accelerate decomposition. Landspreading requires very low inputs of labor and technology and is therefore relatively inexpensive.

Unlike composting, materials that have been landspread are first degraded by the actions of soil-dwelling macroorganisms such as worms and insects. Once the feedstock is size reduced by these macroorganisms, mesophilic microorganisms begin decomposition, which proceeds at low temperatures and slow rates (CRS, 1989). Since the feedstock is applied to the land before any processing is conducted, this method is not appropriate for MSW, which is more likely to contain pathogenic and chemical materials than yard trimmings. Yard trimmings that have been exposed to high pesticide levels also should not be landspread.

To increase the efficiency of the landspreading, the feedstock materials can be shredded prior to application. This increases the uniformity of the particle size of the materials, thereby accelerating composting. Some states govern the level of application of materials to acreage according to water quality concerns and agronomic soil tests. Siting the operations as close to the source of the feedstock materials as possible also should be pursued to minimize transportation costs. For these reasons, careful consideration should be given to siting landspreading operations.

Landspreading of materials that decompose rapidly can enhance plant growth. If the feedstock is applied at the appropriate time, the decomposition process will be completed before crops are planted. The decomposed feedstock materials will then act as a soil amendment product and assist in crop growth. If, however, crops are planted before the decomposition is completed, landspread leaves can reduce crop yield by tying up otherwise available nitrogen and reducing oxygen availability. Also, extensive separation operations might be needed to remove unwanted materials such as brush and glass. Finally, raw leaves and grass can be difficult to handle and have a tendency to clog farm machinery.

facilities that compost large volumes of material. These machines can be either self-propelled or mounted to front-end loaders. Self-propelled windrow turners can straddle windrows, minimizing the required space between windrows and consequently reducing the space requirements for the composting process. Windrow turners should perform several functions including increasing porosity of the pile, redistributing material to enhance process homogeneity, and breaking up clumps to improve product homogeneity.

Aerated Static Piles

Aerated static piles, sometimes called forced aeration windrows, are a relatively high-technology approach that can be used to compost both yard trimmings and MSW. This approach is effective when space is limited and the composting process must be completed within a year. In this method, piles or windrows are placed on top of a grid of perforated pipes. Fans or blowers pump or pull air through the pipes and, consequently, through the composting materials. This maintains aeration in the compost pile, minimizing or eliminating the need for turning. In some operations, the pipes are removed after 10 to 12 weeks of composting and the piles or windrows are then turned periodically.

Aerated static piles are 10 to 12 feet high on average. To facilitate aeration, wood chips (or other porous materials) are spread over the aeration pipes at the base of the pile. The feedstock is then added on top of the wood chips. It might be necessary to top off the pile with a layer of finished compost or bulking agent. This protects the surface of the pile from drying, insulates it from heat loss, discourages flies, and filters ammonia and potential odors generated within the pile (Rynk et al., 1992). It can take as little as 3 to 6 months to produce finished compost with this method.

Air can be supplied to the process through a suction system or a positive pressure system. The suction system draws air into and through the pile. The air then travels through a perforated pipe and is vented through a pile of finished compost, which acts as an odor filter (see Figure 4-11). With this system, condensate from water vapor drawn from the pile must be removed before the air reaches the blower. The ability to contain exhaust gases for odor treatment is an important advantage of suction aeration. The presence of this odor filter, however, more than doubles the pressure losses of suction aeration.

The positive pressure aeration system uses a blower to push air into the compost pile. The air travels through the pile and is vented over its entire surface. Because of the way air is vented, odor treatment is difficult with positive pressure aeration. The absence of an odor filter, however, means lower pressure losses with this system, which results in greater air flow from the same blower power. Therefore, positive pressure systems can be more effective at cooling the pile and are preferred when warm temperatures are a major concern (Rynk et al., 1992).

To ensure that decomposition proceeds at high rates, temperature and oxygen levels must be closely monitored and maintained with aerated static pile composting. Aeration management depends on how the blower is controlled. The blower can be run continuously or intermittently. Continuous operation of the blower permits lower air flow rates because oxygen and cooling are supplied constantly; however, this leads to less uniform pile temperatures. Intermittent operation of the blower is achieved with a

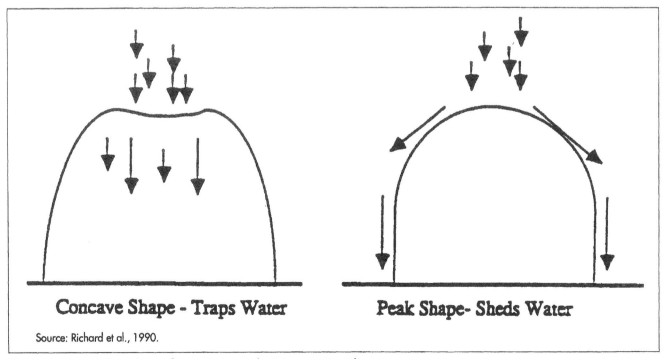

Source: Richard et al., 1990.

Figure 4-10. Windrow shapes for maximum and minimum water adsorption.

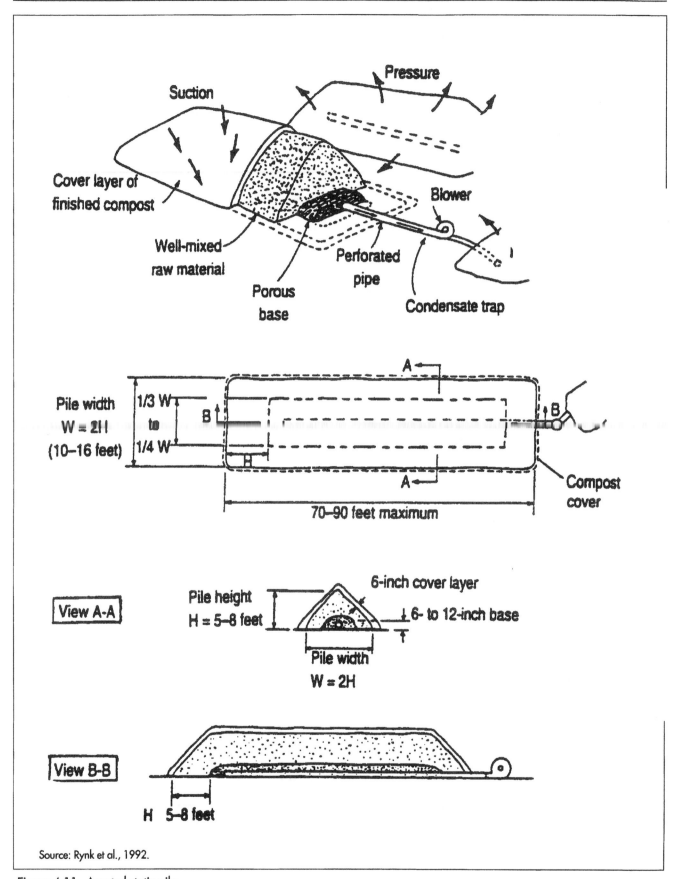

Figure 4-11. Aerated static pile.

programmed timer or a temperature feedback system. Timers are a simple and inexpensive method of controlling blowers to provide enough air to satisfy oxygen requirements and control temperatures. This approach does not always maintain optimum temperatures, however. A temperature feedback system does attempt to maintain optimum pile temperatures, for example, within the range of 54 to 60°C (129 to 140°F) (Rynk et al., 1992). Electronic temperature sensors, such as thermocouples or thermistors, switch the blower on or off when the temperature exceeds or falls below a predetermined level. The blower switches on to provide cooling when the temperature rises above its high temperature, usually around 57°C (135°F), and switches off when the pile cools below a set point (Rynk et al., 1992).

In general, the aerated static pile method is best suited for granular and relatively dry feedstock materials that have a relatively uniform particle size of less than 1.5 to 2 inches in diameter. This is because large or wet materials and materials of diverse sizes have a tendency to clump. Clumping constricts air flow through the piles, leads to short circuits of air pumping equipment, produces anaerobic pockets, and otherwise limits the rate of decomposition. Aerated static piles are commonly used for composting wet materials (such as biosolids), however. Clumping is controlled by proper mixing of bulky materials that adjust porosity and moisture.

In-Vessel Systems

In-vessel systems are high-technology methods in which composting is conducted within a fully enclosed system. All critical environmental conditions are mechanically controlled with this method, and, with most in-vessel systems, they also are fully automated. These systems are rarely used to compost yard trimmings because it is expensive to maintain this degree of control. More and more facilities are selecting in-vessel systems for their MSW composting program. An in-vessel system can be warranted for MSW if: 1) the composting process must be finished rapidly, 2) careful odor and leachate control are a priority, 3) space is limited, and 4) sufficient resources are available.

In-vessel technologies range from relatively simple to extremely complex systems. Two broad categories of in-vessel technologies are available: rotating drum and tank systems. Rotating drum systems rely on a tumbling action to continuously mix the feedstock materials. Figure 4-12 illustrates a rotating drum composter. The drums typically are long cylinders, approximately 9 feet in diameter, which are rotated slowly, usually at less than 10 revolutions per minute (CRS, 1989). Oxygen is forced into the drums through nozzles from exterior air pumping systems. The tumbling of the materials allows oxygen to be maintained at high and relatively uniform levels throughout the drum. The promotional literature for rotating drums indicates that composting materials must be retained in the drums for only 1 to 6 days (CRS, 1989). Complete stabilization of the composting material is not possible within this timeframe, however, and further composting and curing of from 1 to 3 months is necessary (CRS, 1989).

Tank in-vessel systems are available in horizontal or vertical varieties. Rectangular tanks are one type of horizontal in-vessel system. These tanks are long vessels in which aeration is accomplished through the use of external pumps that force air through the perforated bottom of the tanks. Mixing is accomplished by mechanically passing a

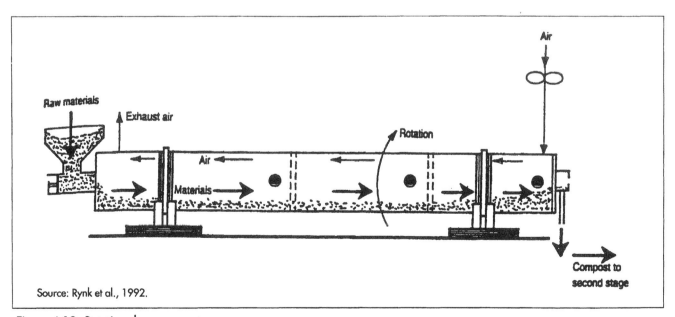

Figure 4-12. Rotating drum composter.

moving belt, paddle wheel, or flail-covered drum through the composting material. This agitates the material, breaks up clumps of particles, and maintains porosity. Composting materials are retained in the system for 6 to 28 days and then cured in windrows for 1 to 2 months.

The agitated-bed system is an example of this type of horizontal in-vessel system. Figure 4-13 illustrates a rectangular agitated-bed composting system. Composting takes place between walls that form long, narrow channels (called beds). A rail or channel on top of each wall supports and guides a compost-turning machine. Feedstock is placed at the front end of the bed by a loader, and the turning machine mixes the composting material and discharges it behind the machine as the material moves forward on rails. An aeration system in the floor of the bed supplies air and cools the composting materials. In commercially available systems, bed widths range from 6 to 20 feet, and bed depths are between 3 and 10 feet. Suggested composting periods for commercial agitated-bed systems range from 2 to 4 weeks (Rynk et al., 1992).

Vertical tank in-vessel systems use a vertical tank orientation. Forced aeration and stirring also are used with this method. These systems can consist of a number of tanks dedicated to distinct stages of the composting process or of one tank (which might be divided into different "floors"). Vertical tank in-vessel systems might use conveyors,

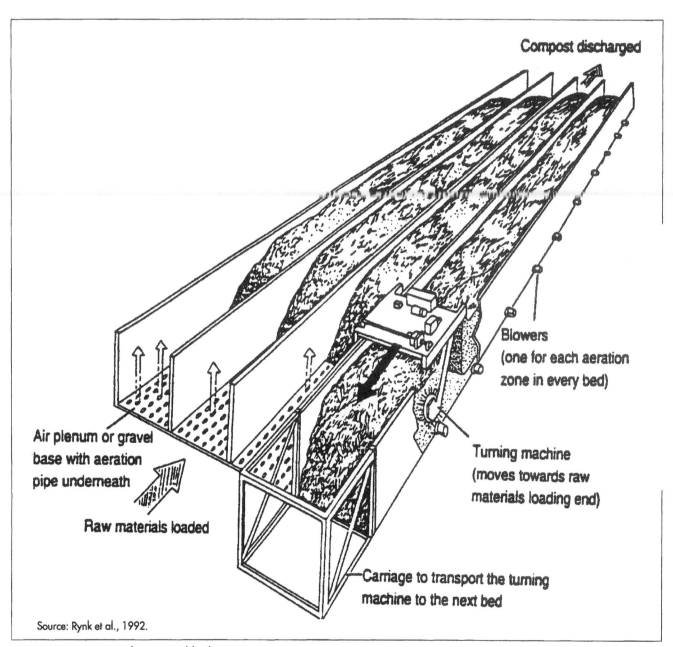

Figure 4-13. Rectangular agitated-bed composting system.

rotating screws, air infeeds, or air outfeeds to agitate compost, move compost between tanks, and maintain proper levels of oxygen and moisture. A problem with vertical tank in-vessel systems is the difficulty of maintaining an equilibrium of moisture and air between the layers inside the tank. In an attempt to adequately aerate the top layers of the compost, these systems can cool down the bottom layers of compost. Furthermore, excessive condensation can form at the top of vertical tanks where moisture and temperature levels are uncontrollable.

The Curing Stage

Once the materials have been composted, they should be cured. Curing should take place once the materials are adequately stable. While testing for stability is an inexact science, oxygen uptake and CO_2 evolution tests can be considered to discern the degree of maturity of compost derived from MSW feedstock. For compost derived from yard trimmings, simpler methods can often suffice. One method is to monitor the internal temperature of the compost pile after it is turned. If reheating of the pile occurs, then the material is not ready for curing. Another method is to put the compost material in a plastic bag for 24 to 48 hours. If foul odors are released when the bag is opened, the materials are not ready for curing.

During the curing stage, compost is stabilized as the remaining available nutrients are metabolized by the microorganisms that are still present. For the duration of the curing stage, therefore, microbial activity diminishes as available nutrients are depleted. This is a relatively passive process when compared to composting stage operations so less intensive methods and operations are used here. In general, materials that have completed the composting stage are formed into piles or windrows and left until the specified curing period has passed. Since curing piles undergo slow decomposition, care must be taken during this period so that these piles do not become anaerobic. Curing piles should be small enough to permit adequate natural air exchange. A maximum pile height of 8 feet often is suggested (Rynk et al., 1992). If compost is intended for high-quality uses, curing piles should be limited to 6 feet in height and 15 to 20 feet in width (Rynk et al., 1992).

Curing operations can be conducted on available sections of the compost storage or processing area. In general, the area needed for the curing process is one-quarter of the size needed during the composting process. The curing process should continue for a minimum of 1 month (Rynk et al., 1992). A curing process of this duration will allow decomposition of the composting materials to be completed and soil-dwelling organisms to colonize the compost. It is important to note, however, that curing is not just a matter of time, it also depends on the favorability of conditions for the process to be completed.

Once the curing process is completed, the finished compost should not have an unpleasant odor. Incompletely cured compost can cause odor problems. In addition, compost that has not been cured completely can have a high C:N ratio, which can tie up otherwise available nitrogen in the soil and be damaging when the compost is used for certain horticultural applications since immature compost can deprive plants of needed oxygen (Rynk et al., 1992). The C:N ratio of finished compost should not be greater than 20:1. C:N ratios that are too low can result in phytotoxins (substances that are toxic to plants) being emitted when composts are used. One group of phytotoxins is produced when excess nitrogen has not been utilized by microorganisms. Nitrogen reactions ultimately can occur, causing the release of ammonia and other chemicals. These chemicals "burn" plant roots and inhibit growth. Therefore, proper end uses for incompletely cured composts are limited (see Chapters 8 and 9).

Odor Control

While odor might seem to be a superficial measure of a composting facility's success, odor is potentially a serious problem at all types of composting facilities and has been responsible for more than one MSW composting plant shutdown. In the planning stage of a facility, decisionmakers should examine composting conditions and odor prevention and control approaches at existing facilities to develop a control strategy for their operations. If nuisance odors still develop, a facility will need to:

- Identify the principal sources of odor.

- Identify the intensity, frequency, characteristics, and meteorological conditions associated with the odors. A facility might consider establishing an "odor standard" above which residents consider the odor a nuisance. An odor panel, made up of community members who volunteer (or are chosen) to represent the community's level of acceptability, can help judge the odor intensity and detectability at their residences.

- Develop limits for odor emissions on site based on maximum allowable odors off site.

- Measure odor release rates from suspected sources for comparison with emission limits.

- Select suitable controls for each source of odor.

Sources of odors include various compounds that may be present in composted organic wastes (such as dimethyl disulfide, ammonia, and hydrogen sulfide). These odors can be produced during different stages of the composting process: conveying, mixing, processing, curing, or storage. Methods exist for measuring the quantity, intensity, pervasiveness, emission rate, and transport of odors and for establishing odor standards. For example, odor quantity can be expressed as the number of effective dilutions (ED) required so that 50 percent of a panel of 10 people can

Case Studies: Odor Problems

Facility managers should anticipate potential odor problems and incorporate odor prevention and control methods from the start. The following are examples of how complaints about odor can lead to setbacks or even failure:

In Illinois, a state law banning yard trimmings from landfills nearly failed when hastily built composting facilities produced unacceptable odor. The Illinois Composting Council was formed to address odor and management issues.

Neighbors of the St. Cloud, Minnesota, MSW composting facility complained about the odors emanating from the facility, resulting in a year-long suspension of large-scale production while the facility constructed an enclosed system and engineered odor controls.

An MSW composting facility in Florida was forced to shut down, partly because of odor complaints. Neighbors would not allow the facility to remain in operation long enough to retrofit the plant and install engineering controls.

---

still detect the odor; this quantity is known as the ED_{50}. Odor standards can be based on odor measurements (e.g., an ED_{10}), the number of odor complaints, or an existing legal standard. Data on relevant meteorological conditions, such as wind speed and direction, temperature, and inversion conditions, often can be obtained from local weather stations. For more information on methods of measuring odors and setting odor standards, see *Control of Composting Odors* (Walker, 1993) and EPA's *Draft Guidelines for Controlling Sewage Sludge Composting Odors* (U.S. EPA, 1992).

The types of odor controls chosen depend on the odor sources, the degree of odor reduction required, and the characteristics of the compounds causing the odor. Odor reduction efforts should incorporate both prevention and control measures. In addition to the process and engineering controls described below, careful monitoring and control of the composting process will help avoid anaerobic conditions and keep odors to a minimum. In-vessel composting tends to cause fewer odor problems, but in-vessel systems still must be operated and monitored carefully. Proper siting (discussed in Chapter 5) and effective public involvement (see Chapter 10) also will help minimize problems resulting from odors.

Process Controls

At facilities that compost yard trimmings, facility managers can implement a number of procedures to minimize odors in the tipping and staging areas. Assuming that grass is cut over the weekend, managers that have control over the collection schedule can arrange for feedstock to be delivered at the beginning of the week to minimize the amount of time that grass is held in closed containers. If grass coming to the facility is already odorous, it should be mixed with a bulking agent (e.g., wood chips) as quickly as possible so that the C:N ratio is approximately 30:1 (Glenn, 1990).

At facilities that compost yard trimmings and/or MSW, procedures that can help prevent or minimize odors include:

- Forming incoming materials into windrows promptly.

- Making sure windrows are small enough to ensure that oxygen can penetrate from the outside and guard against the formation of a foul-smelling anaerobic core but large enough for the interior to reach optimal temperatures. For an aerated pile composting system, the pile height should be limited to 9 feet high (Walker, 1993).

- Providing aeration by completely mixing the feedstock and regularly turning the piles (see Engineering Controls below). Because turning can release odors, however, a windsock can be used for determining when conditions are right for turning so as to keep odors from leaving the site.

- Breaking down piles that are wet and odorous and spreading them for drying. Mixing in dried compost that has been cured also can help.

- Covering compost piles with a roof to help control temperature and moisture levels.

- Avoiding standing pools of water or ponding through proper grading and use of equipment (see Chapters 5 and 6).

Engineering Controls

Facilities that compost yard trimmings typically rely on regular turning of windrows to mitigate odors. Many MSW composting facilities, however, are beginning to use sophisticated odor control technologies to treat exhaust gases from decomposing feedstock. Some facilities collect and treat odorous gases from the tipping and composting areas. Such systems are necessary if simpler odor control measures are unsuccessful. Table 4-3 describes and compares the effectiveness of several odor control methods: odor piles, biofilters, wet scrubbers, adsorption, dispersion enhancement, and combustion. Combustion is effective but can be expensive (Ellis, 1991). Biofilters and air scrubbers, however, are gaining acceptance as effective means for odor control. These two methods are described below.

Table 4-3. Effectiveness of composting odor control technologies.

Technology	Description	Effectiveness
Odor Pile	Odorous gases from composting pile are diverted to flow over finished compost.	Questionable.
Biofilter	Controlled application of odor pile approach, incorporating filter media to which microorganisms are attached.	90%+ removal.

Wet Scrubbers
    Packed tower
    Mist scrubbers
	Odorous compounds are absorbed into a liquid then extracted with chemicals.	Up to 70% per stage.
		< 90%.

Adsorption	Gases are passed over an inert medium to which the odor-causing compounds attach, thereby "cleaning" the gases.	Effective for polishing and control of volatile organic compounds.

Dispersion Enhancement
    Site modification
    Tall stack
	Facilitates greater dispersion of odorous gases.	Moderate.
		Potentially good.

Combustion	Gases are captured and odorous compounds burned.	99%+ removed.

Biofilters

Biofilters have been used to treat odorous compounds and potential air pollutants in a variety of industries. The composting industry is expanding its use of biofilters as engineering design criteria for this technology have become increasingly available (Williams and Miller, 1992a).

In a biofiltration system, a blower or ventilation system collects odorous gases and transports them to the biofilter. The biofilter contains a filtration medium such as finished compost, soil, or sand. The gases are evenly distributed through the medium via a perforated piping system surrounded by gravel or a perforated aeration plenum (an enclosure in which the gas pressure is greater than that outside the enclosure). The incoming gas stream is usually moisturized to keep the filter medium from drying out (Williams and Miller, 1992a).

As the gases filter up through the medium, odors are removed by biological, chemical, and physical processes. Biofilters have an enormous microbial population. For example, soil biofilters contain 1 billion bacteria and 100,000 fungi per gram of soil. These microorganisms oxidize carbon, nitrogen, and sulfur to nonodorous carbon dioxide, nitrogen, sulphate, and water before those compounds can leave the filter medium (Bohn and Bohn, 1987). The biofilter medium acts as a nutrient supply for microorganisms that biooxidize the biodegradable constituents of odorous gases. Biofilters also remove odorous gases through two other mechanisms that occur simultaneously: adsorption and absorption (Naylor et al., 1988; Helmer, 1974). Adsorption is the process by which odorous gases, aerosols, and particulates accumulate onto the surface of the filtering medium particles. Absorption is the process by which odorous gases are dissolved into the moist surface layer of the biofilter particles (Williams and Miller, 1992a). As microorganisms oxidize the odorous gases, adsorptive sites in the filtering medium become available for additional odorous compounds in the gas stream. This makes the process self-sustaining (Williams and Miller, 1992a) and results in long-term odor removal.

Several different biofilter designs have been used in the composting industry. Figure 4-14 illustrates open and enclosed biofilter systems. In an open system, the biofilter is placed directly on the soil surface, or portions can be placed below the soil grade. Typically, an appropriate area of soil is excavated, an aeration pipe distribution network is placed in a bed of washed gravel, and the area is filled with the filter medium. A closed system consists of a vessel constructed of concrete or similar material with a perforated block aeration plenum. The vessel is filled with the biofilter materials.

The type of design chosen depends on the amount of land available, climate, and financial resources. Both open and closed systems can be covered to minimize the effects of precipitation (Williams and Miller, 1992a).

For successful odor control using biofilters, only a few design limitations must be kept in mind:

- The vessel and the medium must be designed to ensure a suitable environment for microbial growth. The moisture content in the biofilter must be optimal for the resident microorganisms to survive and metabolize gases (Williams and Miller, 1992b). It can be very challenging to maintain the proper moisture conditions within the biofilter.

- The biofilter medium must have a large reactive surface area, yet be highly porous. These two characteristics tend to be mutually exclusive in naturally occurring soils and compost; therefore, porous material is often mixed with the soil or compost to

Processing Methods, Technologies, and Odor Control

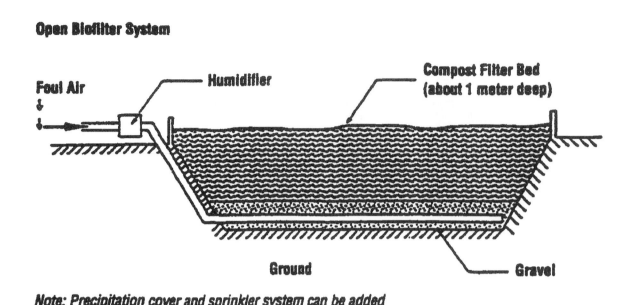

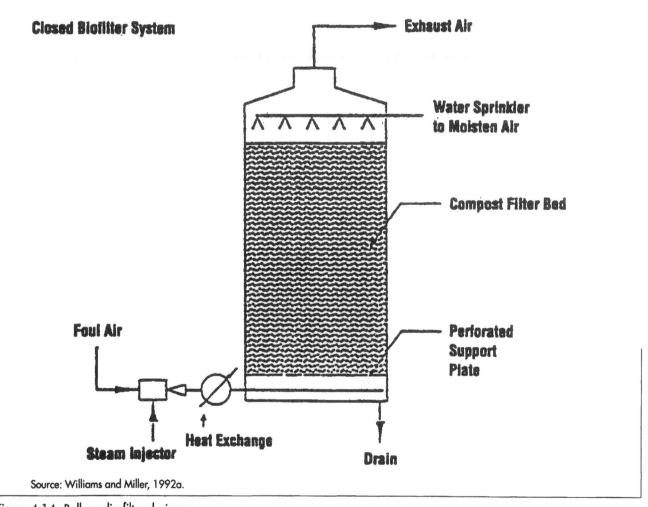

Source: Williams and Miller, 1992a.

Figure 4-14. Bulk media filter designs.

obtain a more suitable biofilter medium (Williams and Miller, 1992a).

- The filtration medium should have a significant pH buffering capacity to prevent acidification from the accumulation of sulfates.

- Compaction of the medium over time should be minimized.

- Uniform air distribution should be designed into the system. If the odorous gases are not distributed evenly throughout the filter medium, "short circuiting" of exhaust gases and inadequate odor control can result (Kissel et al., 1992; Williams and Miller, 1992a).

Table 4-4 presents the maximum removal capacities of various compounds through biofilters. To effectively remove ammonia from composting exhaust gases, other removal technologies such as acid scrubbing (discussed below) might be needed in addition to biofilters.

The initial cost of biofilters is usually less than the installation costs of other odor control methods, and the savings in operation and maintenance are even greater because biofilters require no fuel or chemical input and little maintenance (Bohn and Bohn, 1987). The initial cost of biofilters is $8-10 per cubic foot of air passing through the filter per minute (cfm).

Air Scrubbers

Air scrubbers use scrubbant solutions to remove odorous compounds through absorption and oxidation. A variety of air scrubbers exist. In packed tower systems, the scrubbant solution is divided into slow-moving films that flow over a packing medium. The air stream being treated is usually introduced at the bottom of the packing vessel and flows upward through the medium (Lang and Jager, 1992). The scrubbant solution is recirculated to minimize chemical usage (Ellis, 1991). In mist scrubber systems, the scrubbant solution is atomized into very fine droplets that are dispersed, in a contact chamber, throughout the air stream being treated. Mist scrubbers use a single pass approach: the chemical mist falls to the bottom of the chamber and is continuously drained (Lang and Jager, 1992; Ellis, 1991).

Recent evidence suggests that multiple stages of scrubbers, called multistage scrubbers, often with different chemical solutions, are required to achieve adequate odor removal efficiency (Ellis, 1991). Figure 4-15 illustrates a multistage odor-scrubbing system for a compost operation.

Research by the Washington Suburban Sanitary Commission at the Montgomery County Regional Composting Facility has identified dimethyl disulfide (DMDS) as the primary odorant in air from the composting process (Hentz et al., 1991). This research has led to the development of a three-stage scrubbing process shown to remove 97 percent of the odor in composting exhaust gases. This process involves an acid/surfactant wash in the first stage to remove ammonia and certain organics, a hypochlorite oxidation stage to remove DMDS and other organic sulfides, and a final hydrogen peroxide wash to dechlorinate and further remove organics (Murray, 1991).

Multistage scrubbing systems require effective operation and maintenance procedures to ensure optimum performance. Therefore, before selecting a multistage scrubbing system for odor control, it is important to consider its maintenance requirements in comparison to other odor control technologies.

Postprocessing

Postprocessing is optional but normally is performed to refine the compost product to meet end-use specifications

Table 4-4. Removal capacities of various compounds through biofilters.

Compound	Maximum Removal Rate	Reference
Methylformiate	35.0 g/kg dry media/day.[1]	Van Lith et al., 1990.
Hydrogen Sulfide	5.0 g S/kg dry peat/day.	Cho et al., 1991.
Butylacetate	2.14 g/kg dry peat/day.[1]	Ottengraf, 1986.
Butanol	2.41 g/kg dry peat/day.[1]	Ottengraf, 1986.
N-butanol	2.40 g/kg dry compost/day.[1]	Helmer, 1984.
Ethylacetate	2.03 g/kg dry peat/day.[1]	Ottengraf, 1986.
Toluene	1.58 g/kg dry peat/day.[1]	Ottengraf, 1986.
Methanol	1.35 g/kg dry media/day.[1]	Van Lith et al., 1990.
Methanethiol	0.90 g S/kg dry peat/day.	Cho et al., 1991.
Dimethyl Disulfide	0.68 g S/kg dry peat/day.	Cho et al., 1991.
Dimethylsulfide	0.38 g S/kg dry peat/day.	Cho et al., 1991.
Ammonia	0.16 g N/kg dry peat/day.	Shoda, 1991.

[1] Converted from g/m3/hr, assuming a media bulk density of 40 lb/CF

Source: Williams and Miller, 1992a.

or market requirements. During postprocessing, compost can be analyzed to ensure that stabilization is complete. Compost also can be tested for chemical or pathogenic contamination and tested to determine nutrient levels, cleansed of unwanted material, sorted by size, screened, size reduced, blended with other materials, stored, and/or bagged.

Sorting and removal operations can be conducted to remove any remaining large particles that could lower the quality of the compost or be aesthetically displeasing. Sorting and removal also may be performed to generate composts of uniform size for end uses where such uniformity is important (such as in horticultural applications). The same equipment can be used in both preprocessing and postprocessing, but for composting operations with continual rather than seasonal inputs of feedstock materials, dedicated equipment provides for a more reliable and convenient systems flow. Where size reduction of finished compost particles is desired for aesthetic or marketing reasons, the use of simple shredding mechanisms should suffice.

Proper storage is necessary to maintain the quality of the compost product. The most common storage problem is inadequate drainage controls, causing the compost to become saturated. Overly wet compost can become malodorous and is heavy and difficult to handle. Provision for adequate drainage is essential when storing compost. In general, the storage area should be large enough to hold 25 percent of the compost produced by the facility each year as well as a large supply of bulking agent, if needed (Alexander, 1990).

During postprocessing, compost that will be used as soil amendment should be tested to ensure that it has been properly cured. Compost stability can be assessed by seed germination tests or by analyzing factors that indicate the level of compost maturity. In seed germination tests, sensitive plant species are planted in the compost and in a soil medium. Germination rates for the plants grown in the compost are compared to those grown in the soil and, if the rates are comparable, they show that the compost has been properly stabilized. Laboratory analyses of important compost parameters such as oxygen consumption,

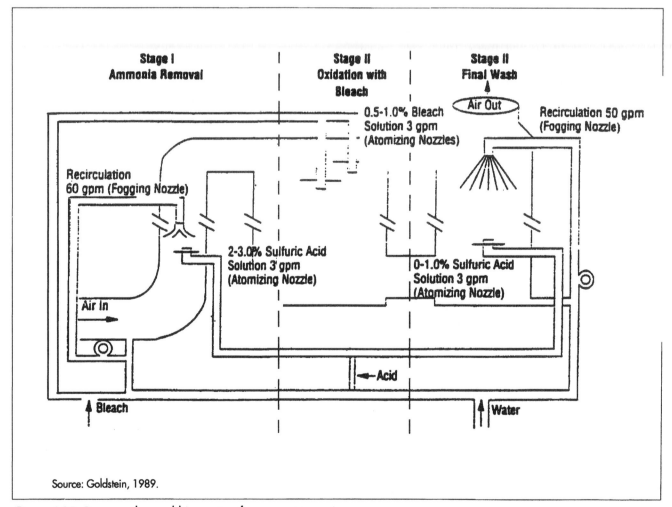

Figure 4-15. Process odor-scrubbing system for compost operation.

carbon dioxide production, C:N ratios, and cation exchange capacity also can be conducted (see Chapter 9).

Laboratory analyses also can be conducted to determine if phytotoxic or pathogenic contaminants are present in the compost. Nutrient levels can be determined through laboratory tests as well. Several states and localities have imposed compost quality requirements (see Chapter 7), and laboratory analysis is often needed to ensure that these requirements are met.

Once contaminant and nutrient levels have been determined, results can be incorporated into compost labels. This will allow end users to obtain composts with contaminant and nutrient levels that fall within ranges acceptable to their specific needs. Labels also can include information on the types of feedstocks used for composting, weight or volume of container contents, suggested uses for the compost, appropriate application rate, warnings or restrictions on compost use, and the name and address of the compost producer.

Finally, compost can be bagged before it is distributed if it is economically feasible. Bagging facilitates transporting, marketing, and labeling of compost. Because it is relatively labor intensive (and therefore costly), however, bagging should be conducted only if buyers for the compost have been secured and the cost of bagging can be justified by an increase in expected revenues.

Summary

There are three stages in the composting process: preprocessing, processing, and postprocessing. Different methods, operations, and equipment are associated with each of these stages. The level of effort applied at each stage depends on the desired quality of the final product, the type and amount of feedstock, the speed at which the process must be completed, the emphasis placed on odor and leachate control, the resources available, and the level of effort applied at the other composting stages. An understanding of the range of methods and operations that can be used during compost processing will facilitate planning and development as well as maintenance and improvement. Composting facility managers also must consider the potential for odor problems when designing processing operations. Odor is a potentially serious problem that has led to the closure of several composting facilities in recent years. Many steps can be taken, however, to address odor formation before it becomes a public nuisance.

Chapter Four Resources

Alexander, R. 1990. Expanding compost markets. BioCycle. August, 31(8):54-59.

Appelhof, M., and J. McNelly. 1988. Yard waste composting guide. Lansing, MI: Michigan Department of Natural Resources.

Bohn, H.L., and R. K. Bohn. 1987. Biofiltration of odors from food and waste processing. As cited in: Proceedings of Food Processing Waste Conference, Georgia Technological Research Institute, Sept. 1-2, 1987.

Buckner, S.C. 1991. High volume yard waste composting. BioCycle. April, 32(4):48-49.

Cal Recovery Systems (CRS) and M.M. Dillon Limited. 1989. Composting: A literature study. Ontario, Canada: Queen's Printer of Ontario.

Composting Council (CC). 1991. Compost facility planning guide. Washington, DC: Composting Council.

Ellis, S. 1991. Air pollution and odor control methods. Proceedings of the Northeast Regional Solid Waste Composting Conference, June 1991. Washington, DC: Composting Council. pp. 23-26.

Glaub, J., L. Diaz, and G. Savage. 1989. Preparing MSW for composting. The BioCycle Guide to Composting Municipal Wastes. Emmaus, PA: The JG Press.

Glenn, J. 1990. Odor control in yard waste composting. BioCycle. November, 31(11):38-40.

Glenn, J. 1991. Upfront processing at MSW composting facilities. BioCycle. November, 32(11):30-33.

Goldstein, N. 1989. New insights into odor control. BioCycle. February, (30)2:58-61.

Golob, B.R., R. Spencer, and M. Selby. 1991. Design elements for solid waste composting. BioCycle. July, 32(7):50-53.

Golueke, C.G. 1977. Biological reclamation of solid wastes. Emmaus, PA: Rodale Press.

Gray, K.R., K. Sherman, and A.J. Biddlestone. 1971a. A review of composting: Part 1 - Process Biochemistry 6(6): 32-36.

Gray, K.R., K. Sherman, and A.J. Biddlestone, 1971b. A review of composting: Part 2 - Process Biochemistry 6(10): 22-28.

Gray, K.R., and A.J. Biddlestone. 1974. Decomposition of urban waste. As cited in: Richard, 1992. Municipal solid waste composting: Physical and biological processing. Biomass & Bioenergy. Tarrytown, NY: Pergamon Press. 3(3-4):195-211.

Processing Yard Waste

The Town of Islip, New York, has been operating a large-scale yard trimmings composting facility on a 40-acre site since 1988. Approximately 60,000 tons of grass, leaves, and wood debris is collected from town residents, municipal agencies, and commercial landscapers every year and transported via packer trucks to the facility.

Islip's composting facility comprises preprocessing, processing, and postprocessing operations. During the preprocessing stage, a shredder debags the composting feedstock and size reduces larger materials. This machine is capable of processing 25 tons of yard waste per hour. Once shredded, the feedstock is conveyed to a trommel screen where it is sorted and aerated. A high percentage of plastic is removed during this preprocessing stage. Once size reduced and screened, moisture is added to the feedstock to obtain an initial moisture content of 50 percent.

During processing, the feedstock is transported via dump trucks to the composting area, where it is formed into windrows on a woodchip base. This base absorbs leachate, increases porosity, and improves drainage conditions at the bottom of the windrows. Twenty-five acres of the facility has been sited for windrow formation. The size of the windrow formed depends upon the nature of the feedstock material and the time of year composting takes place. Feedstock containing mostly leaves can be formed into windrows 12 feet high by 26 feet wide. The size of the windrow formed from feedstock containing predominantly grass depends on the bulking material used, however, these are generally no larger than 6 feet high by 14 feet wide. To maintain aerobic composting conditions, smaller windrows are turned with a rotary-drum turning machine, while a front-end loader is used to turn larger windrows. The frequency of turning varies depending on the windrow size, feedstock composition, stage of decomposition, and moisture content and is adjusted so that aerobic conditions are maintained.

Leaves usually remain in windrows for at least 16 weeks before being placed in curing piles for further stabilization. Grass remains in windrows from 6 to 8 weeks before being placed in curing piles where it will stay for another 4 to 6 weeks. The facility ensures continual processing of fresh material delivered to the site by closely controlling the decomposition rate and windrow size.

Once cured, postprocessing takes place to produce the final product. This involves screening the material to remove woodchips and any plastic fractions remaining in the compost. An air classifier is to be added to the system to separate the plastic from the woodchips so that the chips can be recycled back to the windrows.

The finished compost is available to residents of Islip free of charge and can be purchased by landscape contractors, turf growers, topsoil suppliers, and nurseries for $6 per yard (Buckner, 1991).

Source: Buckner, 1991.

Helmer, R. 1974. Desodorierung von geruchsbeladener abuft in bodenfiltern. Gesundheits-Ingenieur. 95(1):21. As cited in: Williams and Miller, 1992a. Odor control using biofilters, Part I. BioCycle. October, 33(10):72-77.

Hentz, L.H., C.M. Murray, J.L. Thompson, L. Gasner, and J.B. Dunson. 1991. Odor control research at the Montgomery County regional composting facility. Water Pollution Control Federal Journal v. 26, Nov/Dec. As cited in: Murray, 1991. Controlling odor. Proceedings of the 1990 Solid Waste Composting Council Conference, November 1990. Washington, DC: Composting Council. pp. 93-96.

Illinois Department of Energy and Natural Resources (IDENR). 1989. Management strategies for landscape waste. Springfield, IL: Office of Solid Waste and Renewable Resources.

Kissel, J.C., C.H. Henry, and R.B. Harrison. 1992. Potential emissions of volatile and odorous organic compounds from municipal solid waste composting facilities. Biomass & Bioenergy. Tarrytown, NY: Pergamon Press. 3(3-4):181-194.

Lang, M.E., and R.A. Jager. 1992. Odor control for municipal sludge composting. Biocycle. August, 33(8):76-85.

Murray, C.M. 1991. Controlling odor. Proceedings of the 1990 Solid Waste Composting Council Conference, November 1990. Washington, DC: Composting Council. pp. 93-96.

Naylor, L.M., G.A. Kuter, and P.J. Gormsen. 1988. Biofilters for odor control: the scientific basis. Compost Facts. Hampton, NH: International Process Systems, Inc.

Richard, T., N. Dickson, and S. Rowland. 1990. Yard waste management: A planning guide for New York State. Albany, NY: New York State Energy Research and Development Authority, Cornell Cooperative Extension, and New York State Department of Environmental Conservation.

Richard, T.L. 1992. Municipal solid waste composting: Physical and biological processing. Biomass & Bioenergy. Tarrytown, NY: Pergamon Press. 3(3-4):195-211.

Composting Municipal Solid Waste Using Enclosed Aerated Windrows

The MSW composting facility in Wright County, Minnesota has been operational since 1991. This facility processes approximately 165 tons per day (TPD) of MSW, and its composting, curing, and storage areas are sized to accommodate up to 205 TPD. Incoming MSW is weighed and discharged onto the concrete tipping floor of the receiving area where some hand separation of recyclables occurs. The receiving area has a storage capacity of approximately 330 tons.

The preprocessing operations at the facility include screening, handsorting, size reduction, and mechanical sorting. The composting feedstock is transferred from the receiving building to a preprocessing building where it is discharged into a trommel screen equipped with knives to facilitate bag opening. Two conveyors transfer the screened material to the handsorting area. One conveyor transports the fines that pass through the screen openings, and the other transports oversized materials. During this stage, handsorting personnel remove recyclables such as high-density polyethylene and polyethylene terephthalate plastics and aluminum cans. Once handsorted, the feedstock is size reduced by a hammermill located in an explosion-proof enclosure with explosion venting. Following this, the shredded feedstock passes underneath an overhead electromagnet to remove ferrous metals.

At this stage the feedstock material is discharged into a mixing drum and water is added to raise the moisture content to an optimal level. The purpose of the mixing drum is to adjust the moisture content, homogenize the waste stream, and screen oversized and nondegradable material that would inhibit downstream process steps. Three separate feedstock streams are generated by this operation. Material less than 2 inches in size is transported to the composting area, material greater than 2 inches but less than 8 inches undergoes additional shredding and screening, and material greater than 8 inches is disposed of in a sanitary landfill.

Composting takes place in an open-sided covered hangar, sized to contain 12 windrows. Feedstock material is placed in one of two primary windrows formed in the middle of the hangar by a central belt conveyor equipped with a traveling tripper and cross belt conveyor assembly. When one primary windrow has been formed, a windrow turning machine will move through the pile and reposition it to the second row, and from there to the third row, and so on. An aeration system draws air through the primary and secondary windrows. The exhaust air passes through a biofilter for odor control (see Section 6). The facility has an extensive leachate collection system.

The composting feedstock remains in the composting area for approximately 60 days after which it is transferred to a hammermill for further size reduction. A screening drum is then used to separate nondegraded materials from this material. The finished compost is stored on an asphalt pad.

Source: Golob et al., 1991.

Rynk, R., et al. 1992. On-farm composting handbook. Ithaca, NY: Cooperative Extension, Northeast Regional Agricultural Engineering Service.

Strom, P.F., and M.S. Finstein. 1989. Leaf composting manual for New Jersey municipalities. New Brunswick, NJ: Rutgers State University.

University of Connecticut Cooperative Extension Service (UConn CES). 1989. Leaf composting: A guide for municipalities. Hartford, CT: State of Connecticut Department of Environmental Protection, Local Assistance and Program Coordination Unit, Recycling Program.

U.S. EPA. 1992. Draft guidelines for controlling sewage sludge composting odors. Office of Wastewater Enforcement and Compliance, Washington, DC.

Walker, J.M. 1993. Control of composting odors. In: Science and engineering of composting. Hoitink and Keener, eds. Worthington, OH: Renaissance Publications.

Williams T.O., and F.C. Miller. 1992a. Odor control using biofilters, Part I. BioCycle. October, 33(10):72-77.

Williams T.O., and F.C. Miller. 1992b. Biofilters and facility operations, Part II. BioCycle. November, 33(11): 75-78.

Wirth, R. 1989. Introduction to composting. St. Paul, MN: Minnesota Pollution Control Agency.

Chapter Five
Facility Siting and Design

Proper siting and design are prerequisites to establishing safe and effective composting facilities. Decision-makers should take care in selecting a suitable site and developing an appropriate design so as to control both construction costs and operational problems over the life of the facility. This chapter describes factors that should be considered when siting and designing facilities for the composting of MSW or yard trimmings. In general, the primary issues to consider involve odor control (see Chapter 4) and bioaerosol concerns (see Chapter 6). While both types of facilities have similar siting and design requirements, more stringent measures are typically needed at MSW composting facilities. Throughout the siting and design process, it is crucial that the needs of the community be accommodated since public acceptance of a facility is key to its success. Local and state requirements also should be reviewed prior to siting and designing composting facilities. Many states have established specific criteria that composting facilities must address during siting and design. The criteria address many technical concerns, including those related to protecting human health and the environment, and can have an impact on facility location, land use, size, and other considerations. In general, detailed engineering plans typically must be approved by the state environmental protection agency in order to obtain a permit to construct and operate a MSW compost facility. (Chapter 7 discusses state legislation including the specific siting, design, and permitting requirements of several states.)

Siting

Finding a suitable location for a composting facility will help a community achieve its composting goals while avoiding a variety of complications that could slow the composting process. A number of technical, social, economic, and political factors will shape decisions on locating a facility. Some of the major factors in facility siting include:

- Convenient location to minimize hauling distances.
- Assurance of an adequate buffer between the facility and nearby residents.
- Suitable site topography and soil characteristics.
- Sufficient land area for the volume and type of material to be processed.

These factors are described in more detail below. Figure 5-1 presents a site assessment form used in New York State for the composting of yard trimmings. This form is designed to obtain an objective assessment of proposed sites for facilities that compost yard trimmings. Various factors affecting siting are rated from 1 to 5, with 1 being least desirable and 5 being most desirable. These ratings are then added to give a total rating for each site. This rating evaluation makes it easier to choose the most appropriate site for a facility that composts yard trimmings. The same form also could be used for a MSW composting facility.

Location

Potentially suitable locations for composting facilities include areas adjacent to recycling drop-off centers and in the buffer areas of existing or closed landfills, transfer stations, and wastewater treatment plants. Current Federal Aviation Administration (FAA) guidelines prohibit siting any type of solid waste facility, including composting facilities, within 10,000 feet (almost 2 miles) of an airport. This is to prevent birds, which could be attracted to the site by potential food sources, from interfering with airplanes.

Facility Siting and Design

Site Name_____   Date of Inspection_____
Site Location Description_____   Inspected by:_____
_____

This form is designed for use in the field, to obtain an objective assessment of the proposed site. The various "factors" considered at each site receive a rating from 1 to 5, with 1 being <u>least</u> desirable and 5 being <u>most</u> desirable.

<u>FACTORS</u>   <u>RATING</u>   <u>COMMENT</u>

1. Site Preparation Costs
   a) compost area development .... _____   _____
   b) access road construction .... _____   _____
   c) security set-up .... _____   _____

2. Site Characteristics
   a) soil characteristics .... _____   _____
   b) proximity to water; streams, lakes .... _____   _____
   c) slope and topography .... _____   _____
   d) acreage .... _____   _____
   e) drainage .... _____   _____

3. Access by Public Roads .... _____   _____

4. Infrastructure
   a) water .... _____   _____
   b) existing access road .... _____   _____
   c) storage .... _____   _____
   d) telephone .... _____   _____
   e) electric .... _____   _____
   f) scale .... _____   _____

5. Proximity to Homes .... _____   _____

6. Proximity to Town in Need .... _____   _____

7. Regional Site Potential .... _____   _____

Figure 5-1. Yard trimmings site assessment form.

Facility Siting and Design

FACTORS	RATING	COMMENT
8. Land Ownership	_____	_____
9. Environmental Impact		
a) tree removal	_____	_____
b) habitat disturbance	_____	_____
10. Impact on Current Use		
a) visual	_____	_____
b) physical	_____	_____
11. Impact on Future Use		
a) visual	_____	_____
b) physical	_____	_____
12. DEC Criteria (minimum distances)		
a) property line, 50 ft.	_____	_____
b) residence or business, 200 ft.	_____	_____
c) potable water well, 200 ft.	_____	_____
d) surface water supply, 200 ft.	_____	_____
e) drainage swale, 25 ft.	_____	_____
f) water table, 24 inches	_____	_____
TOTAL RATING	_____	

General comments relative to suitability of site to serve as a municipal composting facility:

_____

_____

Source: Richard et al., 1990.

Figure 5-1. (Continued).

A centrally located facility close to the source of the compost feedstock will maximize efficiency and convenience while reducing expenses associated with hauling these materials and distributing the finished compost product. Siting a facility that can be accessed via paved, uncrowded roads through nonresidential areas will further contain transportation expenses. If necessary, however, a busy local road network can be compensated for by scheduling feedstock and compost product deliveries during off-peak road use times. A centrally located facility can offer a further advantage to communities operating drop-off collections since convenient siting often encourages greater resident participation in such programs.

Often, however, the concerns of local residents (particularly about potential odors) force a composting facility to be sited away from ideal collection and distribution locations. This is especially true for MSW composting facilities. Locating a site with an extensive natural buffer zone, planted with trees and shrubs, is an effective way to reduce the potential impacts that a new composting facility might have on the surrounding neighborhoods. If natural buffers do not exist, artificial buffer zones might need to be constructed. Visual screens, such as berms or landscaping, can be designed to protect the aesthetic integrity of the surrounding neighborhoods. (Buffer zones are discussed in more detail later in this chapter.)

Odor Evaluation

A most important consideration in the siting and design of a composting facility is the potential for odors and for odor transport to the community. When planning a facility, it is important to predict potential sources of odors along with their emission rates, detectibility, and intensity. This information can be obtained from literature studies and visits to other composting sites. In order to predict how these odors will be transported, information on meteorological conditions (e.g., wind speed and direction, temperature, and inversion conditions) in the vicinity of the site can be obtained from a local weather station. This information then can be used to conduct dispersion modeling to predict how odors could be transported into the community and how potentially bad they will smell. Data from the modeling can assist decision-makers in choosing a suitable site and in selecting a composting system whose design will help minimize odors (Walker, 1992). (See Chapter 4 for a more in-depth discussion of odor control and management.)

Topography

Potential sites should be evaluated in regard to the amount of alteration that the topography requires. Some clearing and grading will be necessary for proper composting, but minimizing this work is desirable in order to reduce expenses and maintain trees on the perimeter of the site, which act as a buffer. A composting site should be appropriately graded to avoid standing pools of water and runoff. To avoid ponding and erosion, the land slope at a composting site should be at least 1 percent and ideally 2 to 4 percent (Rynk et al., 1992). U.S. Geological Survey topographic maps and a plot plan survey can provide information on the natural drainage characteristics of a site.

The type and structure of the soil present at the site should be assessed to control run-on and runoff. If the site is unpaved, the soil on the site should be permeable enough to ensure that excess water is absorbed during periods of heavy precipitation and that the upper layers of the soil do not become waterlogged (this can create pooling and limit vehicular access). If the soil is impermeable or the site is paved, a range of drainage devices can be used to divert precipitation away from the composting pad and storage areas (see Chapter 6 for more information on these devices).

Proximity to certain water sources also must be considered. Floodplains, wetlands, surface waters, and ground water all need to be shielded from runoff or leachate that can originate at the site. The height of the water table is a crucial factor in protecting these water sources. The water table is the upper surface of the "zone of saturation," which is defined as the area where all available spaces or cracks in the soil and rock are filled with water. In general, the water table should be no higher than 24 inches below the soil surface. Otherwise, flooding can occur during times of heavy precipitation, which can potentially wash away windrows and carry compostable materials off site. Pooling also can result, slowing composting significantly (Richard et al., 1990). In addition, leachate from composting operations is more likely to contaminate ground water when there is less soil to naturally filter the leachate as it seeps downward (Richard, 1990).

Some states have stringent regulations concerning the protection of ground water at a composting site (see Chapter 7). The state of Illinois does not allow the placement of compost within 5 feet of the high water table; North Carolina requires composting pads and storage areas to be at least 2 feet above the seasonal high water table; and Pennsylvania does not allow a composting facility to be sited in an area where the seasonal high water table is less than 4 feet from the surface (WDOE and EPA, 1991).

Flood hazard maps, available from local soil conservation offices, can help show the hydrologic history of a site. In addition, municipalities should research the guidelines that apply in their area as many states have regulations restricting composting operations on floodplains or wetlands. In areas where no local or state regulations exist, Section 404 of the Clean Water Act, administered by the U.S. Army Corps of Engineers, regulates siting issues in proximity to wetlands.

The composting site should have a water source for properly controlling the moisture content of the composting process. The amount and source of the water to be supplied depends on the nature of the compostables, the composting technology used, the size of the operation, and the climate. For example, dry leaves generally require 20 gallons of water per cubic yard of leaves (Richard et al., 1990). Feedstocks with high moisture content (e.g., food scraps) will require less water (see Chapter 2).

Onsite water sources are needed for composting that requires substantial water use. Possible sources include city water hookups, stormwater retention facilities, and wells or surface pumping from nearby lakes or streams. For smaller sites or those requiring minimal amounts of water, mobile water sources can be used. Potential sites should be able to accommodate both the present and future water requirements of the composting program.

Land Area Requirements

To operate efficiently, a composting facility must allot sufficient space to the preprocessing, processing, and postprocessing compost stages as well as to the surrounding buffer zone. Typically, the bulk of the site will be occupied by the composting pad and the buffer zone. (The size of the composting pad and buffer zone are discussed in more detail later in this chapter.) Administrative operations and equipment also need to be housed on site and should be planned for when determining land area requirements for the facility.

Communities should be careful not to locate a facility on too small a site as this can decrease plant efficiency and increase operational costs. The land area of a composting facility must be large enough to handle both present and future projected volumes. Ideally, a composting facility should have, at a minimum, enough acreage to accommodate an entire year's projected volume of incoming feedstock on the site (Richard et al., 1990).

Other Factors Affecting Siting Decisions

Municipalities must consider a number of other factors when siting a composting facility. These factors include:

- *The existing infrastructure* - The presence of existing utility hookups, storage space, and paved access roads could significantly reduce costs of site preparation.
- *Zoning issues* - The construction of composting facilities is permitted only on certain tracts of land within a community as dictated by local zoning laws.
- *Site ownership* - Potential sites could be owned by a public or private entity; ownership will affect cost and control of the composting facility.
- *Nearby land uses* - Sites near schools or residential areas could provoke objections from citizens concerned about potential odor or noise.

Design

Once a site has been identified, a facility must be designed to meet the community's composting needs. It is a good idea to visit other composting facilities to view different designs and operations first-hand. (Figures 5-2 and 5-3 illustrate sample composting site designs.) When developing the initial facility design, future expansion possibilities should be considered in the configuration. Different scenarios should be developed to account for feedstock type and volume changes, facility modifications, system alterations, and other potential revisions in facility design or capability (CC, 1991).

The following are critical to the design of a facility:

- Preprocessing area
- Processing area
- Postprocessing area
- Buffer zone
- Access and onsite roads
- Site facilities and security

Preprocessing Area

A preprocessing or staging area offers room to receive collected feedstock and sort or separate materials as needed. Receiving materials in a preprocessing area will eliminate the need for delivery trucks to unload directly into windrows in poor weather conditions. The size and design of the preprocessing area depends on the amount of incoming materials and the way the materials are collected and sorted (see Chapters 3 and 4). Some facilities also find it advantageous to use a staging area to store separated materials and to wet and hold the materials briefly to prepare them for windrow formation.

The tipping area (the part of the preprocessing area where incoming feedstocks are unloaded) is often roofed in areas subject to severe weather conditions. The floor should be strong enough to support collection vehicles and hardened to withstand the scraping of equipment such as front-end loaders. The tipping floor also should contain no pits, which can attract vermin. Concrete floor slabs and pushwalls to run the front-end loaders against when scooping MSW will increase the efficiency of the operation. The minimum ceiling height of an enclosed tipping area depends on the clearances that the various types of hauling vehicles require to discharge their MSW. The tipping floor area should allow a minimum maneuvering distance of no less than one-and-a-half times the length of the delivery vehicle.

The preprocessing area is also frequently used to shred the compostable material or separate the bags in which the feedstock has been collected. The size of this area depends on the volume of material that the site handles and the sophistication of the system design. For example, the required floor area for a simple system consisting of infeed and discharge conveyors, a single shredder, and a trommel is approximately one-half of that required for a more complex system that also includes vibratory screens, a preshredding flail mill, and postprocessing equipment. A composting site that will sort out recyclables from the MSW received will require additional space and containers for holding these materials.

Some composting facilities use a truck weigh scale to keep track of the weight of feedstock being hauled into the facility as well as the amount of finished compost produced and distributed. Weigh scales of varying lengths can be purchased to accommodate large vehicles. Designed to operate under a variety of weather conditions, they often are located outdoors on the entrance roadway. A scale should be used unless the composting operation is very small.

Processing Area

The processing area, composed of the composting pad and the curing area, must be carefully designed for efficient composting. Design specifications for this area will

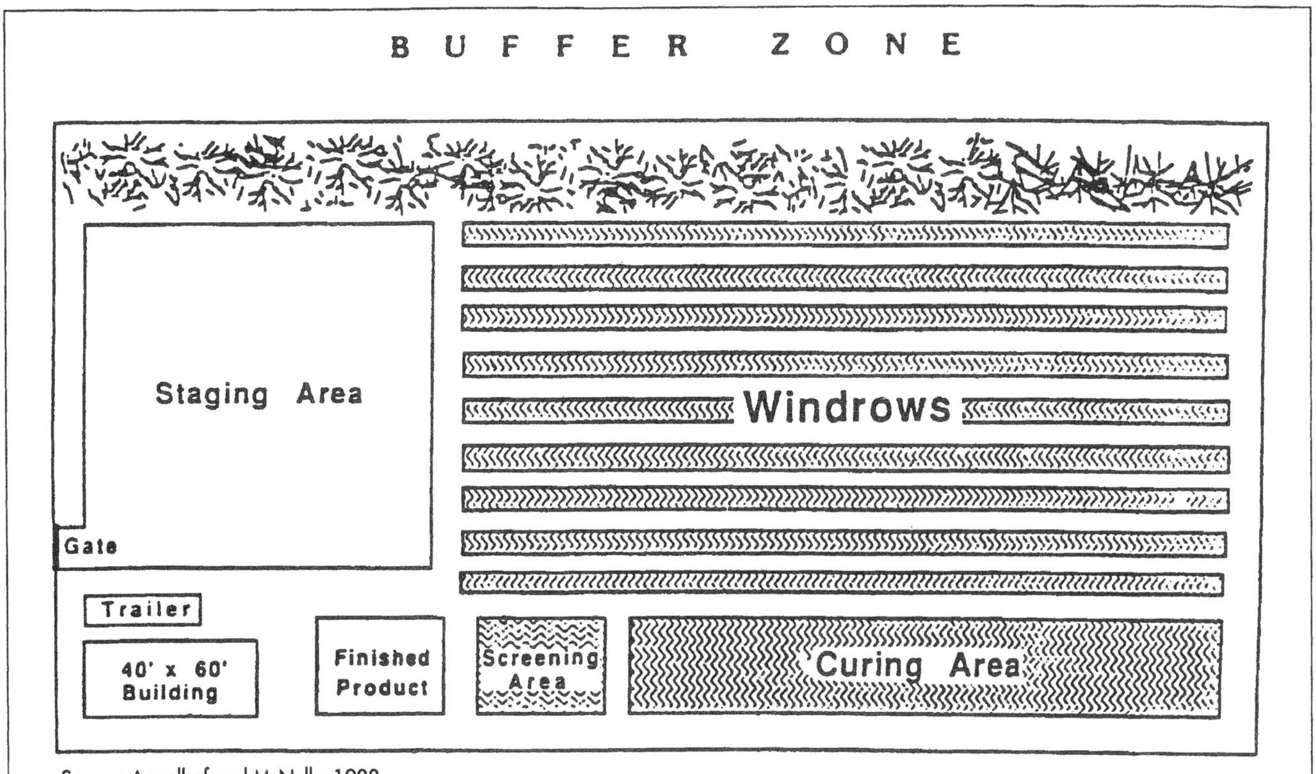

Source: Appelhof and McNelly, 1988.

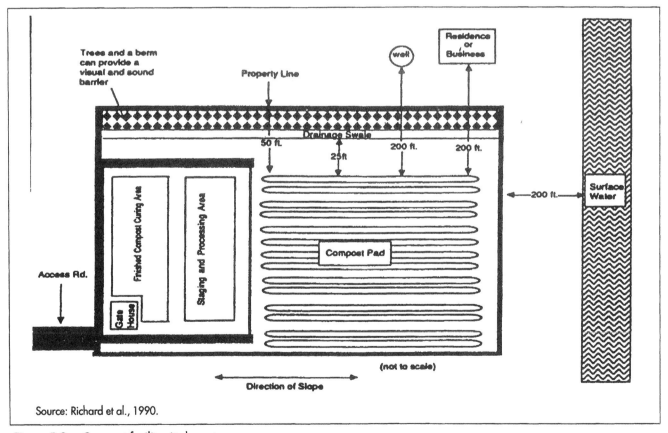

Figure 5-2. Generic compost site layout.

Source: Richard et al., 1990.

Figure 5-3. Compost facility site layout.

differ considerably depending on whether the composting facility processes yard trimmings or MSW feedstocks.

The composting pad surface in a yard trimmings composting facility does not have to be paved; however, it must be firm and absorbent enough to prevent ponding around the windrows or erosion from runoff. Grading the surface of the pad to meet the optimal slope also will help prevent erosion by allowing for gentle drainage. Maintenance of the composting site should include annual re-grading to preserve this slope. As a further protection against erosion, windrows should be arranged parallel to the grade to allow runoff to flow between the piles instead of through them (Richard et al., 1990; Mielke et al., 1989). Precipitation moving onto the composting pads can be diverted from compost piles through the use of drains and conduits. Adequate drainage at composting facilities is essential. Poor site drainage leads to ponding of water, saturated composting materials, muddy and unsightly site conditions, bad odors, and excessive runoff and leachate from the site (Rynk et al., 1992).

Some states have additional requirements for the processing area. For example, to minimize leachate from migrating into subsurface soils, ground water, or surface water, Minnesota requires MSW composting operations to be placed on liners made of synthetic materials, such as high density polyethylene plastics, or natural soils, such as clay. Soil liners must be at least 2 feet thick and compacted to achieve a permeability of no greater than 1×10^{-7} centimeters per second (WDOE and EPA, 1991). Minnesota regulations also require that MSW composting facilities be designed to collect and treat leachate. The preferred method is to collect, pump, and haul the leachate to the municipal wastewater treatment plant if the plant accepts the leachate. Iowa regulations require composting facilities to use an impervious composting pad with a permeability coefficient of 1×10^{-7} centimeters per second (WDOE and EPA, 1991). Florida regulations require MSW composting facilities to conduct their composting operations on surfaces such as concrete and asphalt. They also require a leachate collection system. Municipalities should check with their state to be sure composting pad designs comply with existing guidelines (see Chapter 7).

The size of the composting pad depends primarily on the amount of material that the facility receives for composting and the level of technology that will be used. The required area also depends on the characteristics of the feedstock; the initial and final density of the composting material and the moisture content will affect the amount of material that will fit on the pad. The windrow turning equipment influences aisle width, which in turn influences the size of the composting pad (see Chapter 4). A common design is to line the windrows in pairs 5 feet apart with 15-foot aisles between each pair. This method uses space efficiently but is only possible when straddle-type turning equipment is available (Mielke et al., 1989).

Operations that use a front-end loader to turn the material require individual rows and aisles between the windrows of 15 to 20 feet. Some composting pad areas are housed under structures with movable side walls. In dry climates, where water is scarce or expensive, a roof over the composting area reduces evaporation and process water requirements. In areas of high precipitation, a roof prevents overly wet compost and anaerobic conditions from developing. In regions that experience severe winters, all or part of the composting area can be located within a heated or insulated building to avoid arresting the biological process due to freezing. Because the composting process requires the use of moisture and enclosed composting operations can create extremely damp conditions, wood structures are not recommended unless they are well treated to withstand high moisture levels.

Proper ventilation is required in enclosed preprocessing and processing areas because the air within the structure can be a source of bioaerosols, odors, dust, and excess moisture. Air filters can be used to clean the exhaust air. Biofilters can be used to absorb odor-producing compounds (see Chapter 4). Adequate vents situated over preprocessing equipment can reduce dust and odors, and fans can be used to help disperse nonpervasive odors in the facility.

A curing area also should be part of the design of the processing site. This area is used to hold the compost for the last phase of the composting process, to allow the material to stabilize and cool. The space requirement for curing is based upon the amount of organic material composted, the pile height and spacing, and the length of time that the compost is cured (Rynk et al., 1992). Locating this operation is less problematic than finding a suitable site for the composting pad provided that the composting process has been carried out properly. If this is the case, the material should be fairly stable and many of the runoff, ground-water contamination, and other siting concerns are mitigated. In addition, the curing area needs less space, requiring only about one-quarter of the area of the compost pad (Richard et al., 1990; UConn CES, 1989).

Postprocessing Area

A postprocessing area at composting facilities can be used to conduct quality control testing of compost; to perform screening, size reduction, and blending operations; to compost in preparation for market; and to store the compost. A space about one-fifth the area of the composting pad is sufficient (Richard et al., 1990).

If the finished compost will not be delivered to the end user within a relatively short period of time, the compost should be covered. Otherwise, winds can transport weed seeds into the piles, which can support the growth of unwanted plants and devalue the product. Backup storage and disposal capacity also should be planned for seasonal markets. Cured compost should be stored away from surface water and drainage paths. A storage capacity of at least 3 months should be incorporated into site designs for composting facilities. Cured compost, which is a source of odors in some meteorological conditions, might be better stored away from the site.

Buffer Zone

The buffer zone frequently needs to be several times the size of the composting pad, particularly when the composting operation is adjacent to residential areas or businesses. Enclosed or higher technology facilities might require less of a buffer zone, since many of the operations are by design closely controlled and contained.

During site design, the direction of the prevailing wind (if one exists) should be noted and the buffer zone extended in this direction. This will help minimize the transport of odor and bioaerosols downwind of the facility. Figure 5-4 shows a sample buffer zone design.

In general, the larger the buffer zone, the greater the acceptance of the facility among residents. The buffer zone required by a composting facility depends largely on the type of feedstock being composted and the level of technology (in terms of monitoring and odor control) employed at the facility. State and local regulations frequently require minimal buffer zone sizes or specify the distances that composting operations must be from property lines, residences, or adjacent businesses and from surface water or water supplies (see Chapter 7).

New Jersey regulations recommend a buffer zone for leaf composting facilities of 150 feet (high-level technology, less than one-year cycle) to 1,000 feet (minimal technology, two- to three-year cycles) (WDOE and EPA, 1991). Buffer zone recommendations are wider in New Jersey (from 150 to 1,500 feet) when grass is included in the composting feedstock because of the greater potential for odors. Iowa regulations require MSW composting facilities to be located at least 500 feet from any habitable residence. Table 5-1 lists the minimal separation distances allowed by the State of Wisconsin for facilities that compost yard trimmings or MSW.

Municipalities should check state and local regulations to be sure all applicable guidelines are being incorporated into their buffer zone design. Because odor problems can force a multimillion dollar facility to shut down, communities might extend composting buffer zones beyond the minimum required. (Other steps to control odors are discussed in Chapter 4.)

Access and Onsite Roads

The type and amount of traffic into and out of a facility should be considered in the design process. Traffic at a site is largely dependent on the volume of materials that flows through the facility and the type of collection system in place. For example, operations that compost municipal yard trimmings will involve intensive use of the roads during periods of peak collections. MSW composting operations, on the other hand, will usually receive a more consistent schedule of deliveries. Although an extensive onsite road network usually is not necessary, there should be permanent roads leading to the tipping and storage areas. These access roads should be graveled or paved to handle large vehicles during adverse weather conditions. This surfacing is expensive, however, and the resulting run-on and run-off must be managed to prevent erosion.

If drop-off collections will occur at the facility, the design should accommodate a greater flow of automobile and light truck traffic. A circular traffic flow can accommodate rapid deliveries, effectively reducing congestion. A

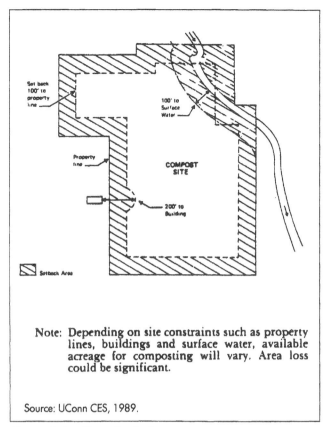

Note: Depending on site constraints such as property lines, buildings and surface water, available acreage for composting will vary. Area loss could be significant.

Source: UConn CES, 1989.

Figure 5-4. Site setback distances.

Table 5-1. Setback requirements for Wisconsin composting facilities.

Navigable lake or pond	1,000 feet
Navigable river or stream	300 feet
State, federal, or interstate highway or public park boundary	1,000 feet
Airport runway	10,000 feet
Public or private water supply well	1,200 feet

Source: WDOE and EPA, 1991.

separate access road to the tipping area also can be constructed for these vehicles (Richard et al., 1990; Strom and Finstein, 1989). Ideally, the road used by the public to deliver materials or to pick up finished compost should be different from the heavy equipment access road. Roads should also be designed to provide adequate turning and dumping areas to accommodate delivery by all types of vehicles.

Site Facilities and Security

Composting operations might require one or more buildings to house various site functions, from maintenance and administrative work to personnel facilities. This is true even for smaller operations such as sites that compost yard trimmings, which might need only a small receiving post. Site buildings should have, at a minimum, electricity, heat, air conditioning, a toilet, and drinking water. All facilities should have a telephone or radio in case of emergencies. In larger facilities (sites with a daily capacity greater than 50 tons), a personnel area containing an office, shower, locker room, and lunch room might be appropriate. A maintenance area that includes a workshop and storage rooms to keep parts and other maintenance materials also might be needed.

Access to the site must be controlled to prevent vandalism, especially arson, and illegal dumping. At a minimum, the access roads must be secured with a fence, cable, locked gate, or other type of constructed barrier. Usually, the surrounding buffer zone will eliminate off-road vehicular access, but if natural geographic barriers do not exist, fencing the entire site might be necessary.

Summary

Today, municipalities face major challenges when attempting to site and design compost processing facilities. When developing a composting facility, municipalities must consider a number of factors, including location, topography, zoning laws, land availability, and ownership. The facility needs to be designed to accommodate both current and projected operations. To ensure that the facility is well sited and designed, input should be sought regarding the technical and economic aspects of a composting system from a range of specialists including engineers, biologists, system managers, and equipment suppliers. Municipalities also must accommodate the needs of local residents throughout the siting and design process to ensure the construction of a facility that the whole community will find acceptable. Community involvement is critical since one of the major factors in the shutdown of many composting operations has been complaints from neighboring households and businesses about odors.

Chapter Five Resources

Albrecht, R. 1992. National Solid Waste Management Association (NSWMA). Seminar on composting. Ft. Lauderdale, FL. November 19-20.

Appelhoff, M., and J. McNelly. 1988. Yard waste composting: Guidebook for Michigan communities. Lansing, MI: Michigan Department of Natural Resources.

Composting Council (CC). 1991. Compost facility planning guide. Washington, DC: Composting Council.

Darcey, S. 1993. Communities put wet-dry separation to the test. World Wastes. 36(98):52-57.

Mielke, G., A. Bonini, D. Havenar, and M. McCann. 1989. Management strategies for landscape waste. Springfield, IL: Illinois Department of Energy and Natural Resources, Office of Solid Waste and Renewable Resources.

Richard, T., N. Dickson, and S. Rowland. 1990. Yard waste management: A planning guide for New York State. Albany, NY: New York State Energy Research and Development Authority, Cornell Cooperative Extension, and New York State Department of Environmental Conservation.

Rynk, R., et al. 1992. On-farm composting handbook. Ithaca, NY: Cooperative Extension, Northeast Regional Agricultural Engineering Service.

Strom, P., and M. Finstein. 1989. Leaf composting manual for New Jersey municipalities. New Brunswick, NJ: New Jersey Department of Environmental Protection, Division of Solid Waste Management, Office of Recycling.

University of Connecticut Cooperative Extension Services (UConn CES). 1989. Leaf composting: A guide for municipalities. Hartford, CT: State of Connecticut Department of Environmental Protection, Local Assistance and Program Coordination Unit, Recycling Program.

Washington Department of Ecology (WDOE) and U.S. Environmental Protection Agency (EPA). 1991. Summary matrix of state compost regulations and guidance. Minneapolis, MN.

Walker, J. 1992. Control of composting odors. As cited in Hoitink, H., and H. Keener, eds. Science and engineering of composting: An international symposium. Worthington, OH: Renaissance Publications. March 27-29, 1992, Columbus, OH.

Chapter Six
The Composting Process: Environmental, Health, and Safety Concerns

> Some aspects of the composting process can pose potential environmental, health, and safety problems. Decision-makers must be aware of these possible complications before proceeding with composting facility planning so that measures can be taken to avert difficulties. This chapter will help officials understand the potential risks involved with composting. Over the past several years, several composting facility closures have occurred due to some of the problems mentioned in this chapter, particularly odor. The first portion of this chapter describes the possible environmental concerns associated with the composting process such as water and air pollution. The second section discusses potential worker health and safety issues. Potential environmental, health, and safety concerns associated with the compost product are discussed in Chapter 9.

Environmental Concerns During Composting

If not carefully controlled, the composting process can create a number of environmental concerns including air and water pollution, odor, noise, vectors, fires, and litter. Many of these concerns can be minimized through the proper design and operation of a facility. In addition, simple procedures often can be implemented to reduce the impact of the facility on the environment.

Water Quality

Water pollution from leachate or runoff is a potential concern at composting facilities. Leachate is liquid that has percolated through the compost pile and that contains extracted, dissolved, or suspended material from the pile. If allowed to run untreated and unchecked from the composting pile, leachate can seep into and pollute ground water and surface water. Runoff is water that flows over surfaces without being absorbed. Contaminated runoff from composting sites can be a problem (particularly at MSW composting facilities) in areas with high rainfall or during periods of heavy rain. Both runoff and leachate also can collect in pools around the facility, producing odor problems. In addition, runoff can cause erosion. There are many ways to prevent and control leachate and runoff at composting operations, as described in the following sections.

Leachate

Leachate from the composting of yard trimmings can have elevated biochemical oxygen demand (BOD) and phenols, resulting from the natural decomposition of organic materials. High BOD depletes the dissolved oxygen of lakes and streams, potentially harming fish and other aquatic life. Naturally occurring phenols are nontoxic but can affect the taste and odor of water supplies if they reach surface water reservoirs. Natural phenols and BOD do not appear to pose a problem to ground water supplies, however, as they are substantially reduced by soil biota through degradation processes (Richard and Chadsey, 1990). Table 6-1 shows elevated levels of phenols and high BOD in leachate from a leaf composting facility in Croton Point, New York.

Another potential water contamination problem at facilities that compost yard trimmings is nitrate generation caused by composting grass clippings along with leaves. Because grass clippings have a low carbon to nitrogen (C:N) ratio, an initial burst of microbial activity depletes oxygen in the composting pile before the grass is completely composted. The lack of oxygen causes aerobic

Table 6-1. Croton Point, New York, yard trimmings compost leachate composition.

	Compost Leachate (16 samples)	
	Average (mg/L)	Standard Deviation (mg/L)
Cd	ND	
Cu	ND	
Ni	ND	
Cr	ND	
Zn	0.11	0.13
Al	0.33	0.38
Fe	0.57	0.78
Pb	0.01	0.02
K	2.70	0.99
NH_4-N	0.44	0.35
NO_3-N	0.96	1.00
NO_2-N	0.02	0.02
Phosphorus	0.07	0.08
Phenols (total)	0.18	0.45
COD	56.33	371.22
BOD	> 41[a]	> 60
pH	7.75	0.36
Color	ND	
Odor	ND	

[a] Includes 3 samples above detection limit of 150 mg/L.
ND = Not Determined.
COD = Chemical Oxygen Demand.
Source: Richard and Chadsey, 1990.

microorganisms to die, releasing nitrates in their cells. One way to avoid nitrate generation is to monitor the C:N ratio, adjusting the feedstock to keep it at optimum levels (see Chapters 2 and 4). At the Croton Point facility (Table 6-1), nitrates were not a problem because grass was not included in the feedstock. Grass clippings can be composted successfully, however, if appropriate material mix ratios, methodology, and equipment are used. In a 3-year study conducted in Massachusetts, very little leaching of nitrate was noted from windrows consisting of one part grass to three parts leaves. Leaching did occur, however, when windrows consisting of grass and leaves in ratios of (or higher than) one part grass to two part leaves were subjected to heavy precipitation or watering (Fulford et al., 1992).

Leachate from yard trimmings and MSW composting operations can also contain potentially toxic synthetic compounds, including polychlorinated biphenyls (PCBs) from treated wood; chlordane, a pesticide; and polycyclic aromatic hydrocarbons (PAHs), combustion products of gasoline, oil, and coal. PCBs and chlordane are resistant to biodegradation and so generally are not broken down during the composting process (Gillett, 1992). While microorganisms can degrade PAHs during the composting process, the compounds formed as a result of this process can be more toxic than the original PAHs (Chaney and Ryan, 1992; Menzer, 1991). Monitoring incoming feedstock to remove pesticide containers and other foreign materials can help reduce the occurrence of synthetic chemicals in leachate.

Leachate generation can be reduced or prevented by monitoring and correcting the moisture levels in the composting pile. In addition, the windrows or piles can be placed under a roof to prevent excessive moisture levels due to precipitation. If the composting materials contain excess moisture, leachate will be released during the first few days of composting even without added moisture or precipitation. Following this initial release of leachate, the amount of leachate formed will decrease as the compost product matures and develops a greater capacity to hold water.

The age of the pile also affects the composition of leachate. As the pile matures, microorganisms break down complex compounds and consume carbon and nitrogen. If the C:N ratio is maintained within the desired range, little excess nitrogen will leach from the pile since the microorganisms will use this element for growth. A study conducted by Cornell University researchers supports this theory (Rymshaw et al., 1992). Table 6-2 summarizes the results of the one portion of the Cornell study in which water was added to columns of manure-bulking agents and the leachates tested for nitrogen content. The leachate produced from 19 weeks of composting and longer was much lower in total nitrogen content than it was in the beginning of the study. Table 6-3 shows concentrations of nitrogen from leachate collected under an actual composting windrow of manure and sawdust. This portion of the study shows an initial peak of nitrogen concentration followed by a subsequent decrease over time. Therefore, as illustrated by this study, the older the composting pile, the less nitrogen will leach from the pile.

Many composting facilities use a concrete pad to collect and control any leachate that is produced (see Chapter 5). The primary task here is to watch the edges, catching any leachate before it leaves the pad. The simplest way to handle leachate is to collect the water and reintroduce it into the compost pile. This should not be done once the composting materials have passed the high-temperature phase, however, as any harmful microorganisms that were inactivated by the high heat can be reintroduced with the leachate (CC, 1991).

Excess amounts of leachate beyond the moisture needs of the composting facility can be transported to a municipal wastewater treatment plant if the plant will accept them. If the plant indicates that the contaminant levels in the leachate are too high, an onsite wastewater pretreatment system might be needed. If leachate is stored, treated on site, or discharged to a municipal wastewater treatment facility, facility operators must comply with federal, state, and local requirements such as regulations covering storage, pretreatment, and discharge permits. It is unlikely that pretreatment will be necessary, however, if the feedstock is monitored carefully. Measures to control leachate include:

- Diverting leachate from the compost curing and storage areas to a leachate holding area.
- Installing liner systems made of low-permeability soils such as clay or synthetic materials.
- Using liners under drain pipes to collect the leachate for treatment.
- Curing and storing compost indoors to eliminate infiltration of leachate into the ground (Wirth, 1989).

Table 6-2. A summary of column study concentrations.

mg/l	Chips/Newspaper Initial/Final	Straw Initial/Final	Sawdust Initial/Final
Nitrate	0.0/13.0	0.0/526.0	7.0/134.0
Ammonia	239.4/11.2	293.1/17.5	800.8/8.71
Organic Nitrogen	975.4/17.5	702.6/25.9	747.3/71.0
Total Nitrogen	1,196.8/28.7	995.7/45.4	1,548.2/79.7
Total Organic Carbon	1,780.8/1,318.1	829.1/1,201.6	1,443.8/995.4
% Water Retained	92.00/85.00	6.67/70.00	781.00/71.25

Laboratory experiment used 10-inch diameter, 24-inch deep columns of manure-bulking agent (woodchips and newspaper, straw, or sawdust) to which water was added. Volumes of water applied corresponded to 2.1 to 12.4 cm of rainfall. Samples were collected from the bottom of the columns over 20 weeks, 21 weeks, or 19 weeks (for chips/newspaper, straw, and sawdust, respectively).

Source: Rymshaw et al., 1992.

Table 6-3. A summary of windrow leachate concentrations.

Weeks	mg/l					
	NO_3	NH_4	Organic Nitrogen	Total Nitrogen	PO_4	Total Organic Carbon
1.0	10.00	28.35	109.90	138.25		8,743.71
1.5	13.00	12.95	115.50	128.45		9,384.00
2.0	10.50	21.00	105.00	126.00		6,258.96
2.5	9.00	25.20	86.80	112.00		5,372.81
3.0	15.00	8.40	134.40	142.80		14,174.92
5.0	3.00	29.80	32.20	62.00		3,715.66
8.0	3.00	39.91		39.91	75.90	
8.5	4.00	14.84	58.80	73.64	50.80	2,459.63

Leachate was collected from under a composting windrow of manure and sawdust.

Source: Rymshaw et al., 1992.

Runoff

Runoff can be caused both by heavy precipitation and by the many aspects of the composting process that use water. For example, the water used to wash trucks and stationary machinery can contribute to runoff. Highly polluted water can be spilled in the tipping area of MSW composting facilities when packer trucks and compaction boxes from restaurants, grocery stores, and food processors are emptied. While MSW facilities are more prone to polluted runoff problems, operations that are composting yard trimmings can also produce runoff containing small quantities of heavy metals, pesticides, and inorganic nutrients.

For both yard trimmings composting facilities and MSW composting facilities, water that has come into contact with incoming raw materials, partially processed materials, or compost should not be allowed to run off the site. Figure 6-1 shows several options for diverting water from composting windrows and for containing runoff from the piles. The facility design must include provisions for isolating, collecting, treating and/or disposing of water that has come in contact with the composting feedstock. These provisions can include:

- Maintaining sealed paving materials in all areas.
- Grading facility areas (1 to 2 percent grade) where contaminated water will be collected.
- Erecting containment barriers or curbing to prevent contaminated water from coming in contact with adjacent land areas and waterways.
- Covering processing areas (composting beds and compost product processing areas).
- Percolating contaminated water through soil so as to absorb and break down organic compounds.
- Creating detention ponds to prevent the discharge of runoff to surface water.

If runoff contains significant amounts of solids (often the case for truck or floor wash-down water), screening, settling, or skimming might be necessary. If runoff is stored, treated on site, or discharged to a municipal wastewater treatment facility, facility operators must comply with federal, state, and local requirements such as regulations covering storage, pretreatment, and discharge permits.

Because runoff can contribute to soil erosion at and around a facility, some simple steps can be taken to avoid soil loss:

- Choosing erosion control measures that are appropriate for the given soil type; more stringent measures are needed for less permeable soil.
- Avoiding sites with steep slopes.
- Grading the site properly (see Chapter 5).
- Minimizing the disruption of existing surfaces and retaining as much vegetation as possible when clearing the site.
- Using proper fill and compaction procedures.
- Prompt seeding and mulching of exposed areas.
- Using erosion screens and hay or straw bales along slopes.
- Using grass filter strips to intercept the horizontal flow of runoff. When runoff passes through the grass strip, pollutants usually settle out of the water or are physically filtered and adsorbed onto the grass.

Run-On/Ponding

Run-on also can be a problem at yard trimmings and MSW composting facilities if the water enters the facility during storms. The site should have a slight slope with windrow piles oriented parallel to the slope to prevent ponding of rainwater among compost piles (Walsh et al., 1990). (See Chapter 5 for more guidance concerning siting and site design.) Ponding or pooling of water on the site also can be a problem if the composting piles rest on a soft surface. Loaders can dig up the dirt base with the piles as they are turned, forming pits that allow water to stand. To remedy this, new fill (e.g., soil, sand, or gravel) should be brought in to replace the excavated material. Equipment that is operated in mud also can create ruts in which ponding can occur. Avoiding work during wet conditions can prevent this problem, although the best way is to compost on paved surfaces.

Air Quality

In general, air pollution is not a major concern at composting facilities, with the exception of the odor problems discussed in the next section. Minor problems could arise, however, from vehicle traffic. The amount of air pollution from vehicle emissions can be reduced by organizing drop-off points to minimize queuing or by restricting feedstock delivery to compaction trucks. Finally, any mobile equipment used at the facility should be well maintained to keep it operating cleanly.

Dust can frequently be a problem at composting facilities, particularly in the dry summer months. Dust is generated from dry, uncontained organic materials, especially during screening and shredding operations, and from vehicle traffic over unimproved surfaces. Dust from composting operations can clog equipment, and carries bacteria and fungi that can affect workers at the facility (see Occupational Health and Safety Concerns During Composting on page 71). As long as there is an adequate buffer zone around the facility, however, residents near the facility

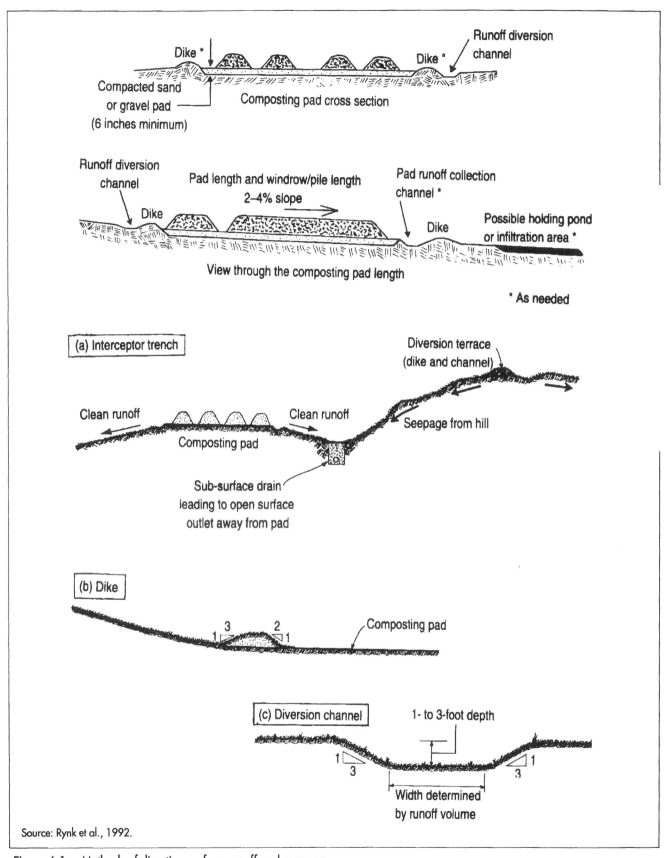

Figure 6-1. Methods of diverting surface runoff and seepage.

Odor

As discussed elsewhere in this manual, odor is a significant concern. Many stages in the composting process can release odors. The feedstock can contain odorous compounds; odors can be produced during collection, transport, and storage of the feedstock or discards; and improper composting procedures can encourage the formation of odorous compounds (Kissel et al., 1992). Anaerobic conditions encourage generation of odorous compounds like organic acids, mercaptans, alcohols, amines, and hydrogen sulfide gas, and other odorous sulfur compounds (Williams and Miller, 1992a; Diaz, 1987). Ammonia can be released under anaerobic conditions and aerobic conditions if the C:N ratio is less than 20:1 (Kissel et al., 1992). The compounds thought to be responsible for odors at composting facilities are listed in Table 6-4. Chapter 4 discusses process and engineering controls for reducing nuisance odors.

Table 6-4. Compounds either specifically identified or implicated in composting odors.

Sulfur Compounds	
Hydrogen Sulfide	Dimethyl Disulfide
Carbon Oxysulfide	Dimethyl Trisulfide
Carbon Disulfide	Methanethiol
Dimethyl Sulfide	Ethanethiol
Ammonia and Nitrogen-Containing Compounds	
Ammonia	Trimethylamine
Aminomethane	3-methylindole (skatole)
Dimethylamine	
Volatile Fatty Acids	
Methanoic (formic)	Butanoic (butyric)
Ethanoic (acetic)	Pentanoic (valeric)
Propanoic (propionic)	3-methylbutanoic (isovaleric)
Ketones	
Propanone (acetone)	2-pentanone (MPK)
Butanone (MEK)	
Other Compounds	
Benzothiazole	Phenol
Ethanal (acetaldehyde)	

Source: Williams and Miller, 1992a.

Noise

Noise is generated by trucks entering and leaving a composting facility and by equipment used in composting operations. Hammermills and other shredding/grinding machines are the noisiest of this equipment, generating about 90 decibels at the source. Many states have noise control regulations that limit noise at the property line.

Measures that can reduce noise emanating from the facility include:

- Providing an adequate buffer zone around the facility with plenty of trees.
- Including specifications for noise-reducing design features, such as mufflers and noise hoods, when procuring equipment.
- Properly maintaining mufflers and other equipment components.
- Coordinating hours of operation with adjacent land uses.
- Taking steps to limit traffic to and from the facility (see "Controlling Air Pollution").

These measures will not always protect workers from exposure to excessive noise on site, however. Further noise control methods are described below under "Occupational Health and Safety."

Vectors

Vectors are small animals or insects that can carry diseases. Mice, rats, flies, and mosquitoes are potential visitors to facilities that compost yard trimmings and/or MSW. Rodents can be attracted by the food and shelter available at composting facilities (particularly MSW composting operations) and can be difficult to eliminate. Where proper operating procedures do not control rodents, the help of a professional exterminator might be required.

Flies, which can transmit salmonella and other food-borne diseases, are often carried in with the incoming material and are attracted to windrows that have become anaerobic. Research has shown that all life stages of the housefly are killed by the temperatures reached in the composting pile (Golueke, 1977). Mosquitoes, which can transmit disease, breed in standing water. Insects can be controlled by keeping the processing area neat, maintaining aerobic conditions and proper temperatures in the windrows, and grading the area properly to prevent ponding.

Fires

If the compost material dries out and becomes too hot, there is a potential for spontaneous combustion to occur at composting facilities. Organic material can ignite spontaneously at a moisture content of between 25 and

45 percent. This is unlikely, however, unless the material reaches temperatures higher than 93°C (199°F), which typically requires a pile over 4 meters high. Keeping the windrows about 3 meters high and turning the compost when temperatures exceed 60°C (140°F) will prevent fires. In addition to these precautions, the site must be designed for access by firefighting equipment, including clear aisles among windrows, and must have an adequate water supply (see Chapter 5) (Richard et al., 1990).

Other steps that can reduce the risk of fire include preventing accumulation of dust produced by shredding operations and preventing in-vessel composting systems from becoming too dry. Adequate site security is necessary to ensure that composting sites do not become a target for arson. Site security will also ensure that the facility does not become a dumping ground for used oil, appliances, or other unacceptable materials.

Litter

Litter from the composting facility can be a source of complaints from nearby residents. Litter can come from yard trimmings and MSW brought to the facility in open loads, plastic and paper blowing from windrows, and rejects (such as plastic) blowing away during screening. Litter can be controlled by:

- Requiring loads of incoming material to be covered.
- Using movable fencing or chain link fences along the site perimeter as wind breaks and to facilitate collection of litter.
- Enclosing receiving, processing, and finishing operations.
- Collecting litter as soon as possible before it becomes scattered off site.
- Removing plastic bags before windrowing or collecting in paper bags, in plastic bins, or in bulk (for leaves and woody materials) (Wirth, 1989).

Occupational Health and Safety Concerns During Composting

Potential health and safety problems at facilities for composting yard trimmings and MSW include exposure to bioaerosols, potential toxic chemicals, and other substances. Excessive noise and injuries from equipment used at the facility also can be concerns. These problems can be minimized by proper siting, design, and operation of the facility and by adequate worker training and education. Additional information about recognizing and controlling job risks can be obtained from Occupational Safety and Health Administration (OSHA) regional offices or from state agencies responsible for occupational health and safety.

Bioaerosols

A variety of biological aerosols (bioaerosols) can be generated during composting. Bioaerosols are suspensions of particles in the air consisting partially or wholly of microorganisms. These microorganisms can remain suspended in the air for long periods of time, retaining viability or infectivity. The bioaerosols of concern during composting include actinomycetes, bacteria, viruses, molds, and fungi. *Aspergillus fumigatus* is a very common fungus that is naturally present in decaying organic matter. The spores of this fungus can be inhaled or can enter the body through cuts and abrasions in the skin. The fungus is not considered a hazard to healthy individuals. In susceptible individuals, however, it can inhabit the lungs and produce fungal infections. Conditions that predispose individuals to infection by *Aspergillus fumigatus* or other molds and fungi include a weakened immune system, allergies, asthma, diabetes, tuberculosis, a punctured eardrum, the use of some medications such as antibiotics and adrenal cortical hormones, kidney transplants, leukemia, and lymphoma (Epstein and Epstein, 1989; Wirth, 1989; USDA and EPA, 1980). Effects due to *Aspergillus fumigatus* exposure are hard to predict because infection depends on worker susceptibility.

Aspergillus fumigatus often colonizes the incoming material at both yard trimmings and MSW composting facilities, and is readily dispersed from dry and dusty compost piles during and after mechanical agitation. The levels of *Aspergillus fumigatus* decrease rapidly only a short distance from the source or a short time after activity stops (Epstein and Epstein, 1989). Table 6-5 shows levels of *Aspergillus fumigatus* in different areas of a biosolids composting facility in Windsor, Ontario. While these data are not from yard trimmings or MSW composting facilities,

Table 6-5. Levels of *Aspergillus fumigatus* at a sewage biosolids composting facility.

Location	Concentration (CFU/m^2)
Mix Area	110 to 120
Near Tear Down Pile	8 to 24
Compost Pile	12 to 15
Front-End Loader Operations	11 to 79
Periphery of Compost Site	2
Centrifuge Operating Room	38 to 75
Grit Building	2
Pump House	10
Background Level	2

CFU = Colony-forming units.
Source: Epstein and Epstein, 1989.

they do demonstrate the direct relationship between fungus levels and activity levels (Roderique and Roderique, 1990). Similar results have been seen in MSW composting plants in Sweden (Clark et al., 1983).

Another health concern at composting facilities is exposure to endotoxins. Endotoxins are toxins produced within a microorganism and released upon destruction of the cell in which it is produced. They can be carried by airborne dust particles. Table 6-6 shows the levels of endotoxins in composts from various sources (Epstein and Epstein, 1989). The levels of endotoxins in the air at one yard trimmings composting facility ranged from 0.001 to 0.014 mg/m^3 (Roderique and Roderique, 1990).

Because bioaerosols and endotoxins are both carried as dust, dust control measures should be incorporated into the design and operation of the facility. These measures help control worker exposure to and reduce the the risk of disease from these airborne hazards. Several steps can be taken to minimize dust generation at the facility:

- Keeping compost and feedstock moist.
- Moistening compost during the final pile teardown and before being loaded onto vehicles, taking care not to over wet the material, which can produce leachate or runoff.
- Constructing driving surfaces from asphalt or concrete (or water can be applied to roadways to minimize dust) (Roderique and Roderique, 1990).
- Minimizing dust from enclosed operations through engineering controls such as collection hoods, negative air pressure at dust generation points, and baghouse technology. These controls, however, tend to be expensive.
- Isolating workers from spore-dispersing components of the composting process such as mechanical turning (for example, using tractors or front-end loaders with enclosed air-conditioned or heated cabs).

- Using aeration systems instead of mechanical turning.

In addition to these control measures, workers should be informed that disease-producing microorganisms are present in the composting environment and that, although the risk of infection is low in healthy individuals, the following precautions should be adhered to for personal protection:

- Workers should wear dust masks or respirators under dry and dusty conditions, especially when the compost is being turned (charcoal-filled respirators also reduce odor perception).
- Uniforms should be provided to employees, and workers should be instructed to wash hands before meals and breaks and at the end of the work shift.
- Shower facilities should be available, and clean clothing and shoes should be worn home by each employee.
- Cuts and bruises should receive prompt attention to prevent contact with the incoming loads or feedstock.
- If the facility is enclosed, proper ventilation is required.

Individuals with asthma, diabetes, or suppressed immune systems should be advised not to work at a composting facility because of their greater risk of infection.

Potentially Toxic Chemicals

Some volatile organic compounds (VOCs), such as benzene, chloroform, and trichloroethylene can present potential risks to workers at MSW composting facilities (Gillett, 1992). Certain solvents, paints, and cleaners contain VOCs. The combination of forced aeration (or periodic turning in the case of window systems) and elevated temperatures can drive VOCs from the composting material into the surrounding atmosphere, much as the aeration and heating of activated biosolids does. Workers are more likely than compost users to be exposed to VOCs. Modeling suggests that this is because most of the VOCs in the feedstock should volatilize from mechanically aerated composting piles within 1 or 2 days (Kissel et al., 1992). To avoid worker exposure to VOCs in enclosed spaces, adequate ventilation is required. Control technologies developed for odor control also apply to VOC control. While misting scrubbers have been used to control VOCs (Li and Karrell, 1990), biofilter design for removing VOCs is not fully developed, however (Kissel et al., 1992). The best method of controlling VOC emissions is to limit their presence in the feedstock. Limiting MSW composting to residential and high-quality commercial feedstocks, instituting source separation, and implement-

Table 6-6. Comparison of endotoxin levels in composts from various sources.

Source	Levels (ng/g)
Biosolids Compost	3.9-6.3
Cattle Manure Compost	2.3
Sheep Manure Compost	4.9
Leaf Compost	4.5

Source: Epstein and Epstein, 1989.

ing effective household hazardous waste collection and education programs can minimize the amount of VOCs in MSW (see Chapter 3).

More persistent organic compounds also pose a potential threat to workers. Workers can be exposed to polychlorinated biphenyls (PCBs), dioxins, pesticides, and polyaromatic hydrocarbons (PAHs) from the composting feedstock and compost itself, although the extent of exposure varies and is hard to determine (Gillett, 1992). Effects on worker health have not been observed from exposure to metals during composting or from the finished compost itself. Mozzon et al. (1987) found that airborne lead and cadmium concentrations were below levels of concern at MSW processing sites (less than 0.003 mg/m^3). Gillett (1992) suggests that compared to workers' exposure to metals in polluted air and food, exposure to metals in compost can be insignificant.

Noise Control

The best way to prevent health effects from excessive noise is to use engineering controls that reduce worker exposure to noise. Regional OSHA offices can provide information to workers and employers regarding sources and control of noise. To prevent hearing loss, workplace noise levels should not exceed 85 decibels (dB). Table 6-7 shows that noise levels in some areas of yard trimmings or MSW composting facilities can exceed 85 decibels. Composting equipment that creates excessive noise should be avoided. It is often possible to purchase screening plants, shredders, and other equipment that do not necessitate the use of ear protection for workers (Appelhof and McNelly, 1988). Simple design control measures such as lowering the height from which feedstock is dropped into processors, rearranging machinery inside the facility, and installing mufflers, can bring noise levels down. Hearing protection should be provided until noisy equipment is repaired or replaced.

Other Safety Concerns

Safe design and operation of equipment used at the composting facility are essential. For example, specialized windrow turning equipment typically has mixing flails that rotate at high speeds and must be well shielded from human contact. Because stones and other objects can be thrown a long distance from turning equipment, operators must ensure a safe clearance around and behind this equipment. Devices that prevent access to equipment undergoing servicing or maintenance might be necessary since unexpected ignition could cause injury to workers. The potential for shredder explosions is discussed in Chapter 4.

Worker Training

Table 6-7. Reported noise levels in resource recovery plants.

Location	Noise Level (dBA)
Tipping Floor	85 - 90
Shredder Infeed	85 - 90
Primary Shredder	96 - 98
Magnetic Separator	90 - 96
Secondary Shredder	91 - 95
Air Classifier Fan	95 - 120
Shop	78
Control Room	70
Offices	67
Maintenance Laborer	89
Shredder Operator	83
OSHA Hearing Conservation Requirements	85
OSHA 8-hr Standard	90
OSHA 4-hr Standard	95

dBA = A-weighted sound-pressure level.
Adapted from: Robinson, 1986.

Worker training is an essential part of ensuring a safe workplace. The objectives of employee safety and health training are:

- To make workers aware of potential hazards they might encounter.

- To provide the knowledge and skills needed to perform the work with minimal risk to health and safety.

- To make workers aware of the purpose and limitations of safety equipment.

- To ensure that workers can safely avoid or escape from emergencies.

Topics that should be covered in health and safety training include the rights and responsibilities of workers under OSHA and/or state regulations; identification of chemical, physical, and biological risks at the site; safe practices and operating procedures; the role of engineering controls and personal protective equipment in preventing injuries and illnesses; procedures for reporting injuries and illnesses; and procedures for responding to emergencies.

Chapter Six Resources

Summary

Environmental and worker health and safety problems can arise during processing. Environmental problems during composting such as water and air pollution, odor, noise, vectors, fires, and litter can be prevented or minimized through proper facility design and operation. Facility planners and managers must also take steps to ensure a safe workplace by reducing potential exposure to pathogens, hazardous substances in composting feedstocks, and excessive noise; by ensuring that equipment is designed and maintained to prevent injuries; and by providing worker training in safety and health concerns.

Appelhof, M., and J. McNelly. 1988. Yard waste composting: Guidebook for Michigan communities. Lansing, MI: Michigan Department of Natural Resources.

Canarutto, S., G. Petruzzelli, L. Lubrano, and G. Vigna Guidi. 1991. How composting affects heavy metal content. BioCycle. June, 32(6):48-50.

Composting Council (CC). 1991. Compost facility planning guide. Washington, DC: Composting Council.

Chaney, R.L., and J.A. Ryan. 1992. Heavy metals and toxic organic pollutants in MSW composts: Research results on phytoavailability, bioavailability, fate, etc. As cited in: H.A.J. Hoitink et al., eds. Proceedings of the International Composting Research Symposium. In press.

Chaney, R.L. 1991. Land application of composted municipal solid waste: Public health, safety, and environmental issues. As cited in: Proceedings of the Northeast Regional Solid Waste Composting Council Conference, June 24-25, 1991, Albany, NY. Washington, DC.

Cimino, J.A. 1982. Carbon monoxide levels among sanitation workers. Proceedings of the 42nd annual AMA congress on occupational health. Tampa, FL. As cited in: J.A. Cimino and R. Mamtani. Occupational Hazards for New York City Sanitation Workers. Journal of Environmental Health. 50(1):8-12.

Clark, C.S., R. Rylander, and L. Larsson. 1983. Levels of gram-negative bacteria, *Aspergillus fumigatus*, dust, and endotoxins at compost plants. Applied Environmental Microbiology. 45(5):1501-1505.

de Bertoldi, M., F. Zucconi, and M. Civilini. 1988. Temperature, pathogen control and product quality. BioCycle. February, 29(2):43-47.

Diaz, L.F. 1987. Air emissions from compost. BioCycle. August, 28(8):52-53.

Dunovant, V.S. et al. 1986. Volatile organics in the wastewater and airspaces of three wastewater treatment plants. Journal of the Water Pollution Control Federation. Vol. 58.

Epstein, E., and J.I. Epstein. 1989. Public health issues and composting. BioCycle. August, 30(8):50-53.

Fulford, B.R., W. Brinton, and R. DeGregorio. 1992. Composting grass clippings. BioCycle. May, 33(5):40.

Gillett, J.W. 1992. Issues in risk assessment of compost from municipal solid waste: Occupational health and safety, public health, and environmental concerns. Biomass & Bioenergy. Tarrytown, NY: Pergamon Press. 3(3-4):145-162.

Golueke, C.G. 1977. Biological reclamation of solid wastes. Emmaus, PA: Rodale Press.

Gordon, R.T., and W.D. Vining. 1992. Active noise control: A review of the field. American Independent Hygiene Association Journal. 53:721-725.

Kissel, J.C., C.H. Henry, and R.B. Harrison. 1992. Potential emissions of volatile and odorous organic compounds from municipal solid waste composting facilities. Biomass & Bioenergy. Tarrytown, NY: Pergamon Press. 3(3-4):181-194.

Lembke, L.L., and R.N. Kniseley. 1980. Coliforms in aerosols generated by a municipal solid waste recovery system. Applied Environmental Microbiology. 40(5):888-891.

Li, R., and M. Karell. 1990. Technical, economic, and regulatory evaluation of tray dryer solvent emission control alternatives. Environmental Progress 9(2):73-78. As cited in: Kissel, J.C., C.H. Henry, and R.B. Harrison. 1992. Potential emissions of volatile and odorous organic compounds from municipal solid waste composting facilities. Biomass & Bioenergy. Tarrytown, NY: Pergamon Press. 3(3-4):181-194.

Menzer, R.E. 1991. Water and soil pollutants. As cited in: Amdur, M.O., J. Doull, and C.D. Klaassen, eds. Casarett and Doull's Toxicology: The Basic Science of Poisons, 4th ed. New York, NY: Pergamon Press. pp. 872-902.

Mozzon, D., D.A. Brown, and J.W. Smith. 1987. Occupational exposure to airborne dust, respirable quartz and metals arising from refuse handling, burning, and landfilling. Journal of the American Industrial Hygiene Association. 48(2):111-116.

Naylor, L.M., G.A. Kuter, and P.J. Gormsen. 1988. Biofilters for odor control: The scientific basis. Compost Facts. Hampton, NH: International Process Systems, Inc.

Pahren, H.R. 1987. Microorganisms in municipal solid waste and public health implications. Critical reviews in environmental control. Vol. 17(3).

Richard, T., N. Dickson, and S. Rowland. 1990. Yard waste management: A planning guide for New York State. Albany, NY: New York State Energy Research and Development Authority, Cornell Cooperative Extension, and New York State Department of Environmental Conservation.

Richard, T., and M. Chadsey. 1990. Environmental impact of yard waste composting. BioCycle. April, 31(4):42-46.

Robinson, W.D., ed., 1986. The solid waste handbook. New York: John Wiley and Sons.

Roderique, J.O., and D.S. Roderique. 1990. The environmental impacts of yard waste composting. Falls Church, VA: Gershman, Brickner & Bratton, Inc.

Rymshaw, E., M.F. Walter, and T.L. Richard. 1992. Agricultural composting: Environmental monitoring and management practices. Albany, NY: New York State Agriculture and Markets.

Rynk, R., et al. 1992. On-farm composting handbook. Ithaca, NY: Cooperative Extension, Northeast Regional Agricultural Engineering Service.

U.S. Department of Agriculture (USDA) and U.S. Environmental Protection Agency (EPA). 1980. Manual for composting sewage sludge by the Beltsville aerated-pile method. EPA/600-8-80-022. Washington, DC: EPA.

U.S. Environmental Protection Agency (EPA). 1989. Characterization of Products Containing Lead and Cadmium in Municipal Solid Waste in the United States, 1970-2000. EPA/530-SW-89-015B. Washington, DC: Office of Solid Waste and Emergency Response.

Walsh, P., A. Razvi, and P. O'Leary. 1990. Operating a successful composting facility. Waste Age. January, 21(1):100.

Williams, T.O., and F.C. Miller. 1992a. Odor control using biofilters, Part I. BioCycle. October, 33(10):72-77.

Williams, T.O., and F.C. Miller. 1992b. Biofilters and facility operations, Part II. BioCycle. November, 33(11):75-79.

Wirth, R. 1989. Introduction to composting. St. Paul, MN: Minnesota Pollution Control Agency.

Chapter Seven
State Legislation and Incentives

Because of the lead role that states have assumed in regulating composting, this chapter focuses on state activities. State legislation has greatly influenced the development of composting approaches in many areas of the country, and a full understanding of early state legislative activity will confer a broad appreciation of legislation issues throughout the country related to the composting of yard trimmings and MSW. This chapter presents an overview of existing state legislation on both yard trimmings and MSW composting and discusses state incentive programs to stimulate yard trimmings and MSW composting. The chapter discusses permit and siting requirements, facility design and operational standards, product quality criteria, bans on landfilling or combustion of organic materials, recycling goals, requirements directed at local governments to implement composting programs, requirements directed at state agencies, and requirements for the separation of yard trimmings and organics from MSW.

Composting Legislation Overview

Adoption and implementation of composting legislation is a cumbersome process, and the status of composting legislation generally lags behind public and legislative interest in the issue. Very few states have composting laws that have been fully implemented, but many states are in the process of enacting legislation or promulgating regulations. In recent years, a surge in legislative activity concerning recycling and composting has occurred, and more composting legislation can be expected in the near future.

In the absence of specific composting legislation, many states and localities regulate yard trimmings and MSW composting facilities under related environmental statutes. For example, many jurisdictions have already implemented regulations governing the composting of sewage biosolids. These jurisdictions often use these regulations to control yard trimmings and MSW composting and, in addition, usually borrow from EPA and state biosolids composting laws when developing specific legislation for the composting of yard trimmings and MSW. In November 1992, EPA issued 40 CFR Part 503, which pertains to the land application, surface disposal, and combustion of biosolids (sewage sludge). Many of the standards promulgated in this rule can be applicable to MSW compost. Many states, in lieu of specific composting standards for MSW, are using these standards as guidelines or as models for regulations. State water and air pollution control laws, solid waste management laws, and environmental protection laws also can be utilized to regulate composting. Of special relevance is Part 503, which governs land application of biosolids and biosolids composting. In addition, a wide range of local ordinances often are applicable, including zoning and building codes, regulations governing materials that can be landfilled or incinerated, fire codes, and safety regulations.

The use of a wide variety of nonspecific local and state ordinances to manage yard trimmings and MSW composting can create a complex regulatory framework. Because of the benefits that can be accrued from composting (e.g., landfill diversion and production of valuable soil amendment products), some states and localities are seeking to stimulate composting by minimizing this regulatory complexity.

There are notable differences between legislation for MSW and yard trimmings composting. The composting of yard trimmings is much more widespread than MSW composting. Consequently, more states have adopted specific legislation regulating the composting of yard trimmings. In general, however, because the composting of yard trimmings poses fewer problems than MSW

State Legislation and Incentives

composting, requirements for the composting of yard trimmings are less stringent than those developed for MSW. Legislation for the composting of yard trimmings is usually general in scope and applies to operations that handle leaves, grass clippings, brush, or some combination of these materials. Legislation in a few states (such as New Jersey), however, targets specific yard trimmings, such as leaves. State MSW composting legislation generally covers household MSW. When any amount of sewage biosolids is co-composted with other materials such as yard trimmings or mixed MSW, it is regulated under EPA's 40 CFR Part 503 regulations.

Table 7-1 presents a summary of legislation at the state level to encourage or mandate composting; Table 7-2 describes specific state legislation used to regulate yard trimmings and MSW composting. These tables can be found at the end of this chapter. The remainder of this chapter discusses specific examples of state legislation pertaining to yard trimmings and MSW composting.

Permit and Siting Requirements

To date, most states (especially those in the central and western United States) have not established specific permit or siting requirements for facilities that compost yard trimmings. In addition, because of minimal environmental impacts generally associated with the composting of yard trimmings, a few states (Delaware, Michigan, and Pennsylvania) have expressly exempted these facilities from any requirements (CRS, 1989). Other states exempt certain types of composting operations from permit and siting criteria. For example, Florida has exempted backyard composting and normal farm operations from composting regulations (FDER, 1989). Wisconsin does not require permits for operations that compost yard trimmings and that are less than 38 m^3 in size (Wis. Stat., 1987-1988). New York also exempts small operations as well as operations that compost only food scraps or livestock manure (N.Y. Gen. Mun. Law, 1990). Illinois does not require permits for composting operations that are conducted on sites where yard trimmings are generated. Composting operations that compost materials at very low rates and most on-farm composting operations also are exempt from permitting (Ill. Rev. Stat., 1989).

Those states that do have siting and permitting requirements for yard trimmings and MSW composting attempt to minimize the impact of composting operations on surrounding property and residences, ensure appropriate composting operations are conducted, and prevent environmental problems associated with leachate runoff. For example, Illinois prohibits siting of facilities for the composting of yard trimmings within 200 feet of a potable water supply or within 5 feet of a water table, inside the 10-year floodplain, or within 200 feet of any residence. In addition, the legislation states that the location of a composting facility shall "minimize incompatibility with the character of the surrounding area" (Ill. Rev. Stat., 1989). New Jersey legislation requires every Soil Conservation District in the state to develop site plans for leaf composting facilities that are to be constructed within their jurisdictions. These site plans must include any information required by the state's Department of Environmental Protection (N.J. Stat., 1990).

Permitting and siting regulations for MSW composting are usually more stringent than those for yard trimmings. Florida regulates mixed MSW composting facilities to the same degree as solid waste disposal sites. These regulations prohibit siting MSW composting facilities in geologically undesirable areas (such as in open sink holes or gravel pits), within 500 feet of a shallow water supply well, within 200 feet of a water body, in an area subject to flooding, within public view from any major thoroughfare without proper screening, on the right-of-way of a public road, or near an airport (FDER, 1990).

Pennsylvania has also adopted a strict set of standards for permitting and siting MSW composting facilities. In order to receive a permit for MSW composting, plans must be submitted to the state's Environmental Quality Board. These plans must describe facility siting and design; facility access; control of leachate, soil erosion, sedimentation, odor, noise, dust, and litter; alternative management of feedstocks or compost in case processing operations or end-use markets; ground-water monitoring; and revegetation and postclosure land use for the site (Penn. Env. Qual. Board, 1988). Strict siting regulations to prevent contamination of surface or ground-water resources are also included in the Pennsylvania rules. For example, siting a facility within the 100-year floodplain or within 300 feet of "an important wetland" is prohibited (Penn. Env. Qual. Board, 1988).

Facility Design and Operations Standards

Most states have not adopted specific regulations for the design and operation of yard trimmings and MSW composting facilities. The legislation that has been adopted attempts to minimize negative impacts associated with composting and to protect public health and the environment. New Jersey has adopted a relatively extensive set of regulations concerning leaf composting operations. These regulations restrict access to composting facilities; limit the maximum quantity of leaves to be composted per acre; limit windrow size; govern windrow placement; restrict the grade of compost pads; establish a minimum turning frequency for windrows; limit the quantity of compost that can be stored on the site; and require the use of leachate, odor, dust, noise, and fire controls. In addition, representatives from the Soil Conservation Districts are required to conduct annual inspections of leaf composting facilities operating within their jurisdiction to ensure that the facilities are properly managed and maintained. Other states have adopted portions of these regulations for facilities that compost leaves and yard trimmings in general.

Florida regulations are similar to those that have been implemented in New Jersey but include some specific requirements geared toward controlling the potential safety, health, and environmental impacts that might be associated with operations that compost mixed MSW. These requirements include prohibitions on the composting of biohazardous wastes and hazardous wastes, except for small quantities of household hazardous wastes. The Florida regulations also include requirements for temperature monitoring and recordkeeping and specify the following: that appropriate stormwater management systems must be implemented at composting facilities; all-weather access roads to the facility must be provided; detailed signs indicating the name and telephone number of the operating authority, hours of operation, charges, etc., must be posted; and litter control devices must be installed (FDER, 1989). In addition to operational requirements similar to those of Florida, Pennsylvania's regulations require that feedstocks are weighed when received, composting equipment is properly maintained, salvaging of materials is strictly controlled, unloading of feedstocks is conducted in a safe and efficient manner, point and nonpoint source pollution of water resources is prevented, soil erosion and sedimentation does not occur, soil and ground-water monitoring is conducted, and residues from composting operations are "disposed or processed at a permitted facility for municipal or residual waste" (Penn. Env. Qual. Board, 1988).

New Jersey regulates mixed MSW composting under the same rules as sewage biosolids composting. Pathogen contamination is consequently regulated in a strict manner and only three specific methods of mixed MSW composting can be used:

- *Windrow composting* - Under this method, aerobic conditions must be maintained, temperatures within 6 to 8 inches of the surface of the windrow must remain above 55°C (131°F) for at least 15 consecutive days, and the windrow must be turned at least five times during this 15-day period.

- *Aerated static pile* - With this method, the pile must be insulated and temperatures of at least 55°C (131°F) must be maintained for a minimum of 3 consecutive days.

- *In-vessel composting* - In this method, the composting mixture must be maintained at a minimum temperature of 55°C (131°F) for at least 3 consecutive days (N.J. Dept. Env. Prot., 1986).

Product Quality Criteria

A few states have adopted a variety of criteria to classify different grades of compost. Criteria covering yard trimmings and MSW composts have been developed that concern the degree of stabilization, particle size, moisture content, levels of organic vs. inorganic constituents, and contaminant content. Florida's regulations governing compost product quality are some of the most well-developed to date. Under these regulations, finished compost products must be tested by approved methods and information must be recorded on the following parameters: percent moisture content; percent of total dry weight of nitrogen, phosphorus, and potassium; percent organic matter; pH; percent foreign matter; mg/kg dry weight of cadmium, copper, lead, nickel, and zinc; and most probable number of fecal coliform. After testing, the quality of the compost is classified based on the type of feedstocks processed as well as strict specifications concerning the maturity of the product, the foreign matter content, the particle size, and metal concentrations. Seven levels of compost quality have been specified:

- *Type Y composts* use yard trimmings as the only feedstock; are mature or semimature; have fine, medium, or coarse particle size; and have a low foreign matter and metal content.

- *Type YM composts* have the same characteristics as Type Y composts but can also use livestock manure as a feedstock.

- *Type A composts* use MSW as the feedstock, are mature, have a fine particle size, and have a low foreign matter and metal content.

- *Type B composts* use MSW as the feedstock, are mature or semimature, have a fine or medium particle size, have an intermediate foreign matter content, and have low or intermediate metal concentrations.

- *Type C composts* use MSW as the feedstock; are mature or semimature, have fine, medium, or coarse particle size, have high foreign matter content, and have high, intermediate, or low metal concentrations.

- *Type D composts* use MSW as the feedstock; are fresh; have fine, medium, or coarse particle size; have a high foreign matter content; and high, medium, or low levels of metals.

- *Type E composts* use MSW as the feedstock and have very high metal concentrations (FDER, 1989).

Under Florida regulations, distribution of compost Types Y, YM, and A are not restricted. Distribution of Types B or C compost is restricted to commercial, agricultural, institutional, and governmental use. In addition, according to the regulations, if the compost "is used where contact with the general public is likely, such as in a park, only Type B may be used" (Fla. Stat., 1989). Distribution of Type D is restricted to landfills or land reclamation projects with which the general public does not generally come into contact. Finally, Type E composts must be disposed of in a solid waste facility (FDER, 1989). Approaches of this kind to regulate compost quality currently are being pursued by several other states.

Pennsylvania has adopted a case-by-case approach for regulating the quality of MSW compost. The state requires that a chemical analysis of MSW compost products be performed and submitted to the Department of the Environment before sale and distribution of the material. The regulations state that "if the Department determines that the compost has the potential for causing air, water, or land pollution," the compost facility operator will be informed that the compost must be "disposed of at a permitted disposal facility" (Penn. Env. Qual. Board, 1988).

Bans on Landfilling or Combustion

Several states have restricted the use of certain disposal options (particularly landfilling and combustion) for yard trimmings. Usually, legislation of this kind is coupled with state efforts to implement composting programs. Even where no overt state efforts exist to initiate the composting of yard trimmings, however, disposal bans indirectly stimulate the composting of yard trimmings. Currently, 21 states have enacted a disposal ban on yard trimmings or components of yard trimmings. Wisconsin and Iowa, for example, have adopted legislation that bans both the landfilling and combustion of yard trimmings (FDER, 1989; Iowa Adv. Legis. Serv., 1990). Illinois, Florida, Minnesota, and Missouri ban disposal of yard trimmings in landfills (Ill. Rev. Stat., 1989; Fla. Stat., 1989; Minn. Stat., 1990; Mo. Adv. Legis. Serv., 1990). New Jersey has banned disposal of leaves in landfills (N.J. Stat., 1990).

Recycling Goals

Recycling goals have been established at the state or local level in many areas. It is generally not mandated that composting be performed in order to meet these goals; the establishment of such goals, however, enhances the attractiveness of composting to states and localities. Some experts believe that without composting it will be difficult to achieve recycling goals of 20 percent or more. Maine and West Virginia are examples of states that have set recycling goals that specifically mandate the composting of yard trimmings (W.Va. Code Ann., 1990, Me. Rev. Stat., 1989).

Requirements for Local Governments to Implement Composting

Some states do require local governments to implement composting programs. For example, state legislation in Minnesota mandates that local governments develop programs for the composting of yard trimmings as part of their overall recycling strategy (Minn. Stat., 1990). Similarly, New Jersey legislation directs localities to develop programs for collecting and composting leaves (N.J. Stat., 1990).

Requirements for State Agencies to Compost

In several states, legislation requires state agencies to participate in composting. For example, Wisconsin legislation mandates that state agencies comply with the state's 100 percent yard trimmings ban (WI Stat 1.59). Some agencies, like the University of Wisconsin-Stevens Point, have complied by creating onsite composting facilities. Other agencies, like the State Capitol Building, have complied by contracting with existing composting facilities. Wisconsin's Department of Administration is responsible for seeing that agencies meet the 100 percent requirement and that no yard trimmings go to landfills from state agencies. Alabama and New Mexico require state environmental departments to evaluate their state agencies' recycling programs (including the composting of yard trimmings) and develop new programs if necessary (Michie's Code of Ala., 1990; N.M. Ann. Stat.).

Separation Requirement

Another method of stimulating the composting of yard trimmings without directly mandating that it occur is to require that yard trimmings be separated from MSW before they are collected. Household separation of yard trimmings facilitates composting by minimizing the need for intensive sorting and removal procedures during compost preprocessing. Legislation in Delaware requires the state's solid waste authority to consider the separation of yard trimmings for potential recycling programs (Michie's Del. Code Ann.). Iowa legislation directs local governments to require residents to separate yard trimmings. Under this legislation, local governments are also instructed to collect yard trimmings if they normally collect other forms of MSW (Iowa Adv. Legis. Serv., 1990).

Yard Trimmings and MSW Composting Incentives

Several states have opted to stimulate yard trimmings and MSW composting through a variety of incentive programs, whether they also subscribe to legislative mandates. State composting incentives include encouraging localities to implement programs and/or giving them specific authority to do so; providing grants to local governments or private firms to develop composting programs; emphasizing market development for compost products; and developing educational programs on backyard composting.

State Encouragement and Local Authority to Implement Programs

MSW management has traditionally been handled at the local level. Many states have consequently opted to maintain and promote such local control. Some of these states have also passed legislation that clearly communicates their support of composting to local governments, however. For example, legislation in both Florida and North Carolina does not mandate that the composting of yard trimmings occur at the local level, but encourages local governments to recycle yard trimmings (Fla. Stat., 1989; Michie's Gen. Stat. of N.C.). Similarly, Iowa legislation does not require composting at the local level, but directs the state to "assist local communities in the development of collection systems for yard waste . . . and . . . the establishment of local composting facilities" (Iowa Adv. Legis. Service, 1990).

Other states have supported local control by specifically granting local governments the authority to mandate yard trimmings and MSW composting. For example, legislation in New York gives municipalities the authority to adopt laws requiring that materials, including garden and yard trimmings, be "separated into recyclable, reusable or other components" (N.Y. Gen. Mun. Law, 1990). Massachusetts' legislation has a similar clause that applies to MSW which states that local governments may establish recycling programs mandating the separation, collection, and processing of recyclables including "compostable waste" (Mass. Ann. Laws, 1990).

Grants

Many states have sought to promote yard trimmings and MSW composting by providing grants to local governments and private businesses to establish composting facilities. For example, Iowa law authorizes the state to "provide grants to local communities or private individuals" that are establishing recycling facilities, including facilities for the composting of yard trimmings (Iowa Adv. Legis. Serv., 1990). In Minnesota, those entities that develop yard trimmings and MSW composting projects can receive "grant assistance up to 50 percent of the capital cost of the project or $2 million, whichever is less" (Minn. Stat., 1990). The state of Washington provides funds, as available, to local governments submitting a proposal to compost food scraps and yard trimmings (Wash. Rev. Code, 1990).

Procurement

State agencies that work to build roads, control erosion, construct buildings, and maintain land consume large quantities of topsoil and organic materials. Many states have committed to developing markets for yard trimmings and MSW compost by setting procurement policies for these agencies. Procurement policies encourage state agencies to buy compost by (1) requiring that state agencies give preference to compost when making purchase decisions or (2) requiring that a given percentage of a state's topsoil/organic material purchases are purchases of compost.

As of April 1993, agencies in Georgia are required to give preference to compost when purchasing topsoil and organic material. The legislation even specifies that the state of Georgia give preference to compost made from source-separated, nonhazardous organics. Several states require agencies to give preference to compost when it is cost effective to do so. These agencies include Florida, Maine, Minnesota, and North Carolina.

Encouragement of Backyard Composting

Several states are encouraging backyard composting of organics. Legislation in Connecticut requires regional jurisdictions to foster recycling through a variety of mechanisms, including backyard composting. These jurisdictions are directed to develop and then implement recycling plans that will facilitate backyard composting of organics. States such as Massachusetts, New Jersey, New York, and Rhode Island have published manuals and brochures that explain backyard composting operations.

Education Programs

Several states encourage yard trimmings and MSW composting through educational and public awareness programs. Massachusetts has initiated a technical assistance program for yard trimmings and MSW composting. The state conducts hands-on workshops and provides guidance materials on designing and operating municipal compost facilities. In addition, state officials visit compost facilities and potential composting sites to provide expert guidance. In Seattle, Washington, an urban, organic gardening organization, Seattle Tilth Association, trains volunteers to teach other city dwellers how to compost yard trimmings and food scraps. The volunteer educators, called "master composters," are thoroughly trained in basic composting methods, compost biology, system design, and troubleshooting. The Seattle Tilth Association also operates a hotline to answer questions about composting.

Summary

Significant legislative activity has occurred at the state level in recent years on both yard trimmings and MSW composting. States have adopted legislation concerning permitting and siting of compost facilities, compost facility design and operation, compost product quality, landfilling or combustion of organic materials, recycling goals, local government implementation of composting programs, state agency composting policy, and the separation of yard trimmings and other organics from MSW. In addition, many states have promoted yard trimmings and MSW composting through a variety of incentive programs that encourage local development of composting and grant local governments the authority to implement such programs, provide funds to local governments or private firms to develop composting programs, stimulate market development for compost products, encourage backyard composting, and advance educational programs.

Chapter Seven Resources

Cal Recovery Systems (CRS) and M. M. Dillon Limited. 1989. Composting: A literature study. Ontario, Canada: Queen's Printer for Ontario.

Florida Department of Environmental Regulation (FDER). 1990. Florida Administrative Code. Solid Waste Management Facilities. Rule 17-701.

Florida Department of Environmental Regulation (FDER). 1989. Florida Administrative Code. Criteria for the Production and Use of Compost Made from Solid Waste. Rule 17-709.

Florida Statutes (Fla. Stat.). 1989. Title XXIX, Public Health. Chapter 403, Environmental Control. Part IV, Resource Recovery and Management, Fla. Stat. 403.70.

Glenn, J. 1992. Solid waste legislation: The state of garbage in America. BioCycle. May, 33(5):30-37.

Harrison, E.Z., and T.L. Richard. 1992. Municipal solid waste composting: policy and regulation. Biomass & Bioenergy. Tarrytown, NY: Pergamon Press. 3(3-4):127-141.

Illinois Revised Statutes (Ill. Rev. Stat.). 1989. Chapter 111 1/2, Public Health and Safety Environmental Protection Act.

Iowa Advance Legislative Service (Iowa Adv. Legis. Serv.). 1990. Seventy-Third General Assembly. Ia. ALS 2153; 1990 Ia. SF 2153.

Maine Revised Statutes (Me. Rev. Stat.). 1989. Title 38, Waters and Navigation. Chapter 13, Waste Management. Subchapter I, General Provisions, 38 M.R.S. 1302.

Annotated Laws of Massachusetts (Mass. Ann. Laws). 1990. Part I, Administration of the Government, Title VII, Cities, Towns and Districts. Chapter 40, Powers and Duties of Cities and Towns, §8H.

Michie's Code of Alabama (Michie's Code of Ala.). 1990. Title 22, Health, Mental Health and Environmental Control. Subtitle 1, Health and Environmental Control Generally. Chapter 22B, Recycling by State Agencies, Code of AlA. 22-22b-3.

Michie's Delaware Code Annotated (Michie's Del. Code Ann.). Title 7, Conservation. Part VII, Natural Resources. Chapter 64, Delaware Solid Waste Authority. Subchapter II, Recycling and Waste Reduction, 7 Del. C. 6453.

Michie's General Statutes of North Carolina (Michie's Gen. Stat. of N.C.). Chapter 130A, Public Health. Article 9, Solid Materials Management. Part 2A, Nonhazardous Solid Materials Management, N.C. Gen. Stat. 130A-309.

Minnesota Statutes (Minn. Stat.). 1990. Environmental Protection. Materials Management, Ch.115A.

Missouri Advance Legislative Service (Mo. Adv. Legis. Serv.). 1990. 85th General Assembly, Second Regular Session. Conference Committee Substitute for House Committee Substitute for Senate Bill No. 530. 1990 Mo. SB 530.

New Jersey Department of Environmental Protection (N.J. Dept. Env. Prot.), Division of Solid Materials Management. 1986. Compost Permit Requirements, August 1.

New Jersey Statutes (N.J. Stat.). 1990. Title 13, Conservation and Development—Parks and Reservations. Chapter 1E, Solid Materials Management, 13:1E-99.12.

New Mexico Annotated Statutes (N.M. Ann. Stat.). Chapter 74, Environmental Improvement. Article 9, Solid Materials Act, N.M. Stat. Ann. 74-9.

New York General Municipal Law (N.Y. Gen. Mun. Law). 1990. Article 6, Public Health and Safety, NY CLS Gen Mun 120-aa.

Pennsylvania Environmental Quality Board (Penn. Env. Qual. Board). 1988. Municipal Materials Management Regulations. Chapter 281, Composting Facilities. Pennsylvania Bulletin, April 9.

Revised Code of Washington (Wash. Rev. Code). 1990. Title 70, Public Health and Safety. Chapter 70.95, Solid Materials Management—Reduction and Recycling, RCW 70.95.810.

Washington Department of Ecology (WDOE) and the U.S. Environmental Protection Agency (EPA). 1991. Summary Matrix of State Compost Regulations and

Guidance. Prepared for the Focus Group Meeting on Compost Quality and Facility Standards. Minneapolis, MN. November 6-8.

West Virginia Code Annotated (W. Va. Code Ann.). 1990. Chapter 20, Natural Resources. Article 9, County and Regional Solid Materials Authorities, W. Va. Code 20-9-1.

Wisconsin Statutes (Wis. Stat.). 1987-1988. Chapter 144, Water, Sewage, Refuse, Mining and Air Pollution. Subchapter IV, Solid Materials, Hazardous Waste and Refuse, Wis. Stat. 144.79.

Table 7-1. State legislation to encourage or mandate yard trimmings and MSW composting.

State	Solid Waste Management Goals (includes source reduction, recycling, and composting unless otherwise specified; mandated unless otherwise specified)	Yard Trimmings Bans	Procurement
Alabama	25%		
Arkansas	40%	Yes	
California	50%		Funds spent on topsoil/organic materials must be spent on compost (20% by 1993; 40% by 1995).
Connecticut	25% recycling alone, which includes yard trimmings composting.		
Delaware	21% recycling alone, which includes yard trimmings composting; not mandated.		
District of Columbia	45% recycling alone, which includes yard trimmings composting.	Yes	
Florida	30% source reduction and recycling, which includes yard trimmings composting.	Yes	State agencies and local governments must procure compost when price is equivalent, especially for highways, recultivation, and erosion control.
Georgia	25%		State agencies must give preference to compost for all road building, land development, and land maintenance.
Hawaii	59%	Yes	
Illinois	25% recycling alone, which includes yard trimmings composting.		
Indiana	50%	Yes	
Iowa	50%	Yes	State and local agencies are directed to give preference to the use of compost in land maintenance.
Kentucky	25%		State agencies must give preference to compost.
Louisiana	25%		
Maine	50% recycling and composting; not mandated.	Yes	All state agencies and public-funded construction/land maintenance activities will use composted and recycled organic material where economically feasible and environmentally sound.
Maryland	25% recycling alone, which includes yard trimmings composting.	Yes	State agencies and local governments must give preference to compost in any public-funded land maintenance activity.
Massachusetts	21% composting specifically.	Yes	
Michigan	8-12% composting specifically; not mandated.	Yes	
Minnesota		Yes	State agencies must use compost where cost effective.
Mississippi	25%		
Missouri	40%	Yes	
Montana	25%; not mandated.		
Nebraska	25%	Yes	State agencies and local governments must give preference to compost.
Nevada	25%		

Table 7-1. (Continued).

State	Solid Waste Management Goals (includes source reduction, recycling, and composting unless otherwise specified; mandated unless otherwise specified)	Yard Trimmings Bans	Procurement
New Hampshire	40%		
New Jersey	25% recycling, which includes yard trimmings composting but excludes leaf composting as part of the goal.	Yes (leaves only).	State agencies and local governments must give preference to compost for public-funded land maintenance activities.
New Mexico	50%		State and local agencies are directed to give preference to the use of compost in land maintenance.
New York	60%; not mandated.		State agencies must give preference to recycled materials, including compost.
North Carolina	25%	Yes	State agencies and local agencies must give preference to compost if it does not cost more.
North Dakota	40%		
Ohio	25%	Yes	
Oregon	50%		State agencies must purchase compost to the "maximum extent economically feasible".
Pennsylvania	25% recycling alone, which includes yard trimmings composting.	Yes (leaves and brush only).	
Rhode Island	15% recycling alone, which includes yard trimmings composting.		
South Carolina	30%	Yes	State agencies and public funded projects must purchase compost where economically practicable.
South Dakota	50%; not mandated.	Yes	
Tennessee	25%	Yes	
Texas*	40%		
Vermont	40%; not mandated.		
Virginia	25% recycling alone, which includes yard trimmings composting.		
Washington	50%		The State Department of General Administration must spend at least 25% of its budget on compost products for use as landscape materials and soil amendments; in July 1994, this figure will be raised to 50%; for contracts that use soil cover on state and local rights-of-way, compost products must comprise 25% of the materials purchased; in July 1993, this figure will rise to 50% for state roads and in July 1994, the figure will rise to 50% for local roads.
West Virginia	50%	Yes	Agencies and instrumentalities of the state must use compost in all landscaping and land maintenance activities.
Wisconsin		Yes	

*The Department of Health must compost 15% of the state's solid waste stream by 1994.
Sources: Glenn, 1992; WDOE and EPA, 1991.

Table 7-2. Legislation to regulate the composting of yard trimmings and MSW.

State	Requirements for Operating Compost Facilities	Requirements for Designing/Siting Compost Facilities	Compost Classification/Quality Standards
California	Must be promulgated by the Department of General Services.		
Connecticut	Specific requirements for leaf composting; leaf composting is exempt from solid waste permitting requirements.		
Delaware			Contaminant Standards.
Florida	Separate requirements exist for yard trimmings and MSW composting.	Same requirements exist for yard trimmings and MSW composting, but additional restrictions are placed on MSW composting.	Compost must be classified based on type of material composted, maturity of compost, foreign matter content, particle size, and heavy metal content; restrictions on use of certain categories of compost exist; compost testing required.
Illinois	Requirements exist for operating landscape composting operations.	Design and siting criteria exist for landscape composting.	
Iowa	Separate, extensive requirements for yard trimmings and MSW composting.	Design criteria for MSW composting; siting criteria for both yard trimmings and MSW composting (slightly more restrictive for MSW composting).	Pathogen control standards for finished compost; finished compost must be innocuous and free of sharp-edged objects.
Maine	Specific requirements for "vegetative trimmings" composting; general requirements apply to biosolids composting, co-composting and composting of vegetative trimmings.	Requirements for buffer distances and impermeable ground surfaces.	Classifications for animal manures and vegetative trimmings.
Massachusetts	MSW composting subject to state solid waste management regulations (as for landfills); guidance is available for yard trimmings composting.	MSW composting, same requirements as landfills; separate requirements and guidelines for facilities that compost yard trimmings.	Requirements for MSW compost; guidelines for yard trimmings compost.
Minnesota	Separate requirements for yard trimmings and MSW composting.	Requirements for MSW composting but not yard trimmings composting.	Classification standards and use restrictions for some categories.
Missouri	Minimal requirements for yard trimmings composting.	Siting criteria for facilities that compost yard trimmings.	Only yard trimmings can be used in the compost; compost from yard trimmings is regulated as fertilizer or soil conditioner, depending on how it is labeled.
New Jersey	Criteria for yard trimmings composting; separate criteria for leaf composting alone.	Requirements and guidelines for yard trimmings composting.	
New York	Facilities that compost yard trimmings or MSW must comply with regulations for solid waste management facilities and must also comply with specific yard trimmings and MSW composting requirements.	Separate yard trimmings and MSW facility design requirements; siting criteria for solid waste management facilities.	Classification requirements for MSW composting but not for yard trimmings.
North Carolina	Specific requirements for operating MSW composting facilities; also requirements for yard trimmings, agricultural, and silvicultural composting.	Specific requirements for designing and siting MSW composting facilities; also siting and design requirements for yard trimmings, agricultural, and silvicultural composting.	Specific classification system and quality standards for MSW compost; also requirements for yard trimmings, agricultural, and silvicultural compost.

Table 7-2. (Continued).

State	Requirements for Operating Compost Facilities	Requirements for Designing/Siting Compost Facilities	Compost Classification/Quality Standards
Pennsylvania	General requirements exist for operating all compost facilities, but these are targeted at biosolids composting; specific requirements exist for operations that compost yard trimmings.	General requirements exist for siting and designing all compost facilities, but these are targeted at biosolids composting; specific siting requirements exist for yard trimmings composting but no specific requirements for facility design.	No compost classification system or quality standards exist for MSW composting, although some standards exist for biosolids composting and the biosolids can be composted with MSW; a classification scheme for yard trimmings compost is in place.
South Carolina			Quality standards for yard trimmings compost are being promulgated.
Virginia	Separation operations criteria exist for yard trimmings and MSW composting, and no mixing is allowed.	Separate design and siting criteria exist for yard trimmings and MSW composting.	Yard trimmings and MSW must be composted separately; MSW may be composted with biosolids.
Wisconsin	Separate operations criteria exist for yard trimmings composting on a small scale, yard trimmings composting on a large scale (over 20,000 cubic yards annually), MSW composting, and biosolids and livestock manure composting.	Siting criteria exist for all composting operations; in addition, operations that compost yard trimmings and are 20,000 cubic yards per year must submit a design plan to the state.	Composting is classified into the following categories: household composting, neighborhood composting, community yard trimmings composting, solid waste composting, and biosolids and livestock manure composting; yard trimmings compost may be used without a permit, but a permit is required to landspread MSW compost.

Sources: WDOE and EPA, 1991; Harrison and Richard, 1992.

Chapter Eight
Potential End Users

Compost currently is used in a variety of applications in the United States, from agriculture and landscaping to reforestation projects and residential gardening. When planning a composting facility, decision-makers should identify potential end users to determine the type of compost that is required. For example, if a facility only produces low-quality unscreened compost and end users demand high-quality screened compost, the product might not be used. By identifying potential end users, a facility can ensure that the compost product can be marketed. MSW composting facilities and large yard trimmings composting facilities might identify markets for a spectrum of end uses, from low- to high-quality compost, since the finished product does not always meet the desired specifications. This chapter discusses the potential end users of compost derived from MSW or yard trimmings.

The Benefits of Finished Compost

Compost can benefit the biological, chemical, and physical properties of soil. Biologically, compost enhances the development of fauna and microflora, renders plants less vulnerable to attack by parasites, and promotes faster root development of plants. Chemically, compost benefits soil in a number of ways. Compost increases nutrient content, turns mineral substances in soil into forms available to plants, and regulates the addition of minerals to soil, particularly nitrogenous compounds. In addition, compost serves as a buffer in making minerals available to plants and provides a source of micronutrients. Furthermore, compost improves many physical properties of the soil, including the soil's texture, water retention capacity, infiltration, resistance to wind and water erosion, aeration capacity, and structural and temperature stability. Table 8-1 summarizes potential end users and their quality requirements for finished compost. These end-use markets are examined in more detail in this chapter.

Agricultural Industry

A market assessment was conducted in 1991 to estimate the potential demand for compost in the United States (Slivka, 1992). This survey identified agriculture as the largest potential end-use market for compost, accounting for over 85 percent of potential use. At present, however, the amount of compost used in large-scale agricultural applications is small. According to a 1992 Composting Council survey of 126 yard trimmings and 20 MSW composting programs, only four yard trimmings and three MSW facilities mentioned the agricultural sector as an end-use market.

Agricultural use of compost remains low for several reasons. One, the product is weighty and bulky, which can make transportation expensive. The nutrient value of compost is low compared to fertilizers. In addition, agricultural users might have concerns regarding potential levels of heavy metals (particularly lead) and other possible contaminants in compost, particularly mixed MSW compost (see Chapter 9). The potential for contamination becomes an important issue when compost is used on food crops. This concern is mitigated if compost is applied well in advance of planting. Many experiments examining the effects of MSW compost application on the physiochemical characteristics of soils have indicated positive results as outlined in Table 8-2 (Shiralipour et al., 1992).

To successfully market a compost product to the agricultural sector, therefore, the compost must be available at the appropriate time of year, be consistent in composition and nutrient content, contain low levels of potentially toxic substances, and be offered at a low cost. Additionally, difficulties associated with bulkiness must be resolved, distribution channels established, and the positive effect of compost on crop yields demonstrated (EPA, 1993).

If these issues are addressed, compost has the potential to be used in large quantities by the agricultural industry. Compost can be used to increase the organic matter, tilth, and fertility of agricultural soils. Compost also improves the aeration and drainage of dense soils, enhances the water-holding capacity and aggregation of sandy soils, and increases the soil's cation exchange capacity (i.e., its ability to absorb nutrients) (Rynk et al., 1992). In addition, compost enhances soil porosity, improves resistance to erosion, improves storage and release of nutrients, and strengthens disease suppression (EPA, 1993). The near-neutral pH of compost also is beneficial for growing most agricultural crops.

An important potential use of compost in the agricultural industry is its application as a soil amendment to eroded soils. Farmers in the United States are increasingly concerned about the depletion of organic matter in soil and are acutely aware that fertility is dependent upon maintaining a sufficient amount of organic matter in the soil (EPA, 1993). Compost is an excellent source of organic matter that can enrich soil and add biological diversity. When applied to eroded soils, compost can help to restore both organic content and the soil structure (Kashmanian et al., 1990).

The use of compost can help restore and build up nutrients in soil. The nutrients in compost are released slowly to the roots of plants through microbial activity over an extended period of time, thereby reducing the potential for nutrients to leach from the soil. The gradual dissipation of nutrients from compost also indicates that only a fraction of the nitrogen and phosphorus available in compost is available to the crop in the first year. When applied continuously, the supply of plant nutrients from compost is enough to keep plants healthy for several years. Studies on the residual properties of compost on agricultural soils have reported measurable benefits for 8 years or more after the initial application (Rynk et al., 1992).

Effects of Compost Application on Crop Yields in Johnson City, Tennessee

Compost has been demonstrated to improve crop yields. A study was conducted in Johnson City, Tennessee, from 1968 to 1972 that involved applying compost made from mixed MSW to test plots. During the period of the study, 13 successful corn crops were produced and yield increases due to compost application were noted. The total increase in yield ranged from 55 percent with an application rate of 40 tons of compost per acre to 153 percent with an application rate of 1,000 tons per acre (Mays and Giordano, 1989). The figure below outlines the increases in crop yields following compost application over a 14-year period.

Effects of MSW Compost Application on Test Plots

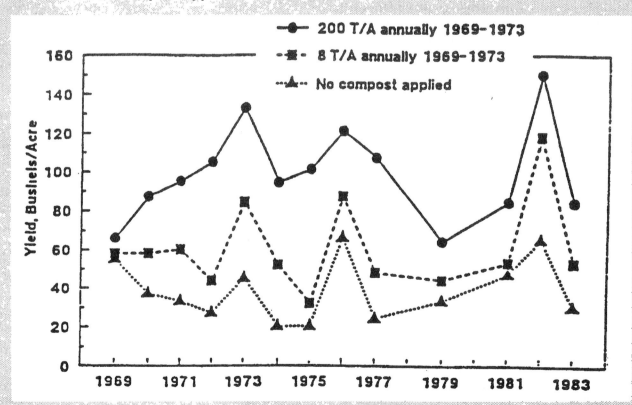

Source: Mays and Giordano, 1989.

Table 8-1. Potential users and uses of compost.

User Group	Primary Uses for Compost Products	Compost Products[a]
Agricultural and residential		
Forage and field-crop growers	Soil amendment, fertilizer supplement, top dressing for pasture and hay crop maintenance	Unscreened and screened compost
Fruit and vegetable farmers	Soil amendment, fertilizer supplement, mulch for fruit trees	Unscreened and screened compost
Homeowners	Soil amendment, mulch, fertilizer supplement, and fertilizer replacement for home gardens and lawns	Screened compost, high-nutrient compost, mulch
Organic farmers	Fertilizer substitute, soil amendment	Unscreened and screened compost, high-nutrient compost
Turf growers	Soil amendment for turf establishment, top dressing	Screened compost, topsoil blend
Commercial		
Cemeteries	Top dressing for turf, soil amendment for turf establishment and landscape plantings	Screened compost
Discount stores, supermarkets	Resale to homeowners	General screened compost product
Garden centers, hardware/lumber outlets	Resale to homeowners and small-volume users	Screened compost, mulch
Golf courses	Top dressing for turf, soil amendment for greens and tee construction, landscape plantings	Screened compost, topsoil blend
Greenhouses	Potting mix component, peat substitute, soil amendment for beds	High-quality, dry, screened compost
Land-reclamation contractors	Topsoil and soil amendment for disturbed landscapes (mines, urban renovation)	Unscreened compost, topsoil blend
Landscapers and land developers	Topsoil substitute, mulch, soil amendment, fertilizer supplement	Screened compost, topsoil blend, mulch
Nurseries	Soil amendment and soil replacement for field-grown stock, mulch, container mix component, resale to retail and landscape clients	Unscreened and screened compost, composted bark, mulch
Municipal		
Landfills	Landfill cover material, primarily final cover	Unscreened low-quality compost
Public works departments	Topsoil for road and construction work, soil amendment and mulch for landscape plantings	Unscreened and screened compost, topsoil blend
Schools, park and recreation departments	Topsoil, top dressing for turf and ball fields, soil amendment and mulch for landscape plantings	Screened compost, topsoil blend, mulch

Note: Unscreened compost with a consistent texture and few large particles may be used in place of screened compost.
[a] Topsoil blend is a mixture of compost, soil, or sand to make a product with qualities similar to topsoil or loam. Mulch includes unscreened, coarse-textured compost such as composted wood chips or bark.
Source: Rynk et al., 1992.

Table 8-2. Effect of MSW compost application on physiochemical characteristics of soil.

Compost Type	Compost Rate (mt/ha)	Duration of Experiment	Experimental Condition	Soil Type	Changes in Soil Physiochemical Characteristics	Reference
MSW ± N-P-K fertilizer	35, 70	2 years	Field	Phosphate mine sand tailings	Increased C.E.C.[a], E.C.[b], O.M.[c], and K, Ca, Mg levels.	Hortenstine and Rothwell, 1972
MSW ± N	4.4, 44	2 years	Field	Sandy soil	Increased water holding capacity, pH, O.M., C.E.C., exchangeable Ca, Mg, and K.	Bengtson and Cornette, 1973
MSW	37–99	3 years	Field	Clay	Made the heavy soil more friable, promoted a crumbly structure, permitted better moisture absorption, reduced erosion, improved aeration, and increased pH.	Duggan, 1973
Pelletized MSW, N-P-K fertilizer	8, 16, 32, 64	16 months	Greenhouse	Arredondo sand	Increased water holding capacity, C.E.C., N, P, K, Ca, Mg, B, Mn, and Zn levels.	Hortenstine and Rothwell, 1972
MSW + SS	80, 112, 143	2 years	Field	Songo silt loam, clay loam	Increased water holding capacity, O.M., pH, and K, Ca, Mg, Zn levels. Decreased bulk density and compression strength.	Mays et al., 1973
MSW ± N (ammonium nitrate)	22, 44, 80, 160, 325	2 years	Field	Songo silt loam, clay loam	Increased water holding capacity, O.M., pH, and K, Ca, Mg, Zn levels. Decreased bulk density and compression strength.	Terman and Mays, 1973
MSW + SS + N	124, 248, 496	5 years	Field	Halston loam	Increased K, Ca, Mg, Zn, but decreased P levels in the soil.	Duggan and Wiles, 1976
MSW	—	—	Field.	Redish-brown clay	Increased O.M. and water holding capacity.	Wang, 1977
MSW	112, 224, 448	5 years	Field	Myakka-Basinger fine sand	About 50% of the applied inorganic P was converted to organic P and remained in the zone of compost placement. Metals were distributed in 0 to 23 cm depth.	Fiskell and Pritchett, 1980
MSW + SS	—	—	Field (mulching)	Sandy loam	Increased pH and P, K, Ca, Mg levels. Reduced erosion.	Sanderson, 1980.
MSW + SS	—	3 years	Field	Alluvial, loamy	Increased C, N, P, K levels.	del Zan et al., 1987
MSW + SS	—	—	Greenhouse	Loamy sand	Increased pH, O.M., and Cd, Cu, Mn, Pb, Zn levels.	Chu and Wong, 1987
MSW	25	—	Field and climate controlled pots	Loamy and sand/vermiculite mixture	Did not excessively increase the heavy metals. Increased pH.	Bauduin et al., 1987
MSW	6, 15, 40	3 years	Field.	—	Increased pH, O.M., and total N.	Paris et al., 1987
MSW	15, 30	2 years, 6 months	Field, greenhouse	—	Increased macro and micro elements, pH, E.C., and O.M.	Manios and Syminis, 1988
MSW, MSW + SS	—	24 years	Field	Luvisol derived from Loess	Increased total C and total N.	Werner et al., 1988

Table 8-2. (Continued).

Compost Type	Compost Rate (mt/ha)	Duration of Experiment	Experimental Condition	Soil Type	Changes in Soil Physiochemical Characteristics	Reference
MSW + SS + N-P-K fertilizer	98.8 to 2,470	5 years	Field	Alluvial, loamy	Increased O.M., C.E.C., pH, macro and micro elements, and heavy metals. Decreased bulk density.	Mays and Giordano, 1989
MSW	14	—	Field	Alluvial	No change in available P levels, but increased K, and available levels of Cu, Zn.	Cabrera et al., 1989
MSW	15, 30, 60	7, 90, 180 days	In chambers	Typic Haploxeralt	Increased aggregate stability, water holding capacity, and pH. No change in C.E.C.	Hernando et al., 1989

Source: Shiralipour et al., 1992.

[a]C.E.C. = cation exchange capacity
[b]E.C. = electrical conductivity
[c]O.M. = organic matter content

Applying compost to soils reduces the likelihood of plant diseases. This is due to several factors. First, the high temperatures that result from the composting process kill pathogens and weed seeds. The frequent turning of windrows and the insulating layers in static piles ensure uniform high temperature exposure and, therefore, uniform pathogen reduction. Second, beneficial microorganisms in compost kill, inhibit, or simply compete with pathogens in soil, thereby suppressing some types of plant disease caused by soil-borne plant pathogens and reducing the need to apply fungicides or pesticides to crops. Microorganisms use the available nutrients in compost to support their activity. Organic matter within compost can replenish the nutrients in soil that has low microbial activity and, as a result, is susceptible to developing soil-borne diseases. Finally, physical and chemical characteristics such as particle size, pH, and nitrogen content also influence disease suppression. Research indicates that some composts, particularly those prepared from tree barks, release chemicals that inhibit some plant pathogens (Hoitink and Fahy, 1986; Hoitink et al., 1991).

Another potential use of compost in the agricultural industry is the prevention of soil erosion. Soil erosion has a direct financial impact on food production and the economy. Composting is one of the few methods available for quickly creating a soil-like material that can help mitigate this loss. Soil erosion also has a serious impact on the quality of the nation's surface water supply. Agricultural runoff from croplands, pasture lands, rangelands, and livestock operations is estimated to be responsible for over 50 percent of the nonpoint source-related impacts to lakes and rivers (Kashmanian et al., 1990). Encouraging farmers to use compost made on and off the farm can both reduce erosion and improve water quality. Some counties in Tennessee and Minnesota are allowed to "cost-share" the agricultural use of compost. The state helps farmers in these areas defray the cost of purchasing or transporting the compost (Kashmanian et al., 1990).

Landscaping Industry

The landscaping industry is another potential outlet for compost. According to the Composting Council survey, the majority of composting facilities surveyed sell compost to landscapers (79 yard trimmings facilities and 12 MSW facilities market to this industry). Landscapers use compost in direct soil incorporation, in the production of outdoor growing mixes, in the manufacture of topsoil for new planting, as a soil amendment, and in turf establishment and maintenance projects. Landscapers require a premium compost. In general, this means that the product should have minimal odor, particle sizes of no greater than 1/2 inch in diameter, less than 50 percent moisture content, and no plant or human pathogens (see Table 8-3). Compost with a near-neutral pH is most suitable for this industry. Every effort should be made, therefore, to avoid using liming or acidifying agents during composting. Landscapers must have the flexibility of raising or lowering pH themselves so the compost can be useful for growing plants with different pH requirements.

The landscaping industry also requires that the materials used in its projects meet the specifications of the landscape architect or inspector. Therefore, compost marketed to this sector must be demonstrated to meet these specifications. Since landscapers also have expressed concern about the possible presence of potentially toxic compounds in MSW compost and of viable seeds, herbicides, and pesticides in yard trimmings compost, tests should be conducted on the final compost product and the results made available to potential users.

The amount of compost used by the landscaping industry depends on economic cycles in the construction and housing industries. For example, new construction projects such as residential housing developments and commercial buildings can create a high demand for compost. The amount of compost used by landscapers also is affected by price, availability, and ease of compost application (EPA, 1993).

Landscapers have successfully used compost as a top dressing to reduce weed growth and improve the appearance of soil and as a mulch to reduce evaporation and inhibit weed growth. Compost is used in the manufacture of topsoil due to its ability to improve the quality of existing soil, which is beneficial to new planting. This use of compost is attractive to landscapers because it can reduce the amount of new topsoil needed, thereby reducing costs.

Other uses of compost in the landscaping industry include maintenance of lawns and parks, highway landscaping, and sod production. Athletic field maintenance, renovation, and construction are other strong potential uses for compost in this industry (Alexander, 1991). Compost can be used as a soil amendment in the renovation of athletic fields, as a turf topdressing to help maintain the quality of the turf surface, and as a component of athletic field mixes, which are used in the construction of new fields.

Horticultural Industry

The horticultural industry is one of the largest potential markets for compost of uniform consistent high quality. Compost is attractive to the horticultural industry because it is a source of organic matter and essential trace plant nutrients, increases the water-holding capacity of soil, improves the texture of soil, and enhances a soil's ability to suppress plant diseases. The use of compost in potting mixtures and in seedling beds has helped to reduce the need to apply soil fungicides in the production of certain horticultural crops (Rynk et al., 1992).

The use of compost by the horticultural industry depends upon the quality of the compost, the consistency and availability, and the cost. As is the case with landscapers, the use of compost in this industry also depends upon the state of the economy, particularly the housing industry. The number of new single-family dwellings built and the number of homes sold have a direct impact on the demand for horticultural products. When home sales rise, the demand for nursery products increases as well (EPA, 1993).

The products distributed to the horticultural industry must be of the highest quality and almost always must be unlimed. Because of its higher pH, limed compost has fewer applications than unlimed compost (Gouin, 1989). To improve the quality of compost earmarked for the horticultural industry, the compost should be thoroughly stabilized. Composting in smaller piles and for longer periods of time aid the stabilization process. It is also impor-

Using Yard Trimmings Compost in Landscaping

Montgomery County, Maryland, sells most of its compost to landscapers and nurseries in minimum loads of 10 cubic yards (EPA, 1993). The facility screens its finished compost, which is derived from leaves and grass clippings, and tests it for weed seeds and heavy metals. Montgomery County has found the peak market demand for its finished product occurs in the spring and fall.

tant for the compost to be odor free. This can be achieved by ensuring that the compost does not become anaerobic during curing or storage. In addition, the compost should be stored either under cover or outdoors in low windrows not to exceed 6 feet in height. Table 8-4 outlines compost quality guidelines based on certain horticultural end uses. These suggested guidelines have received support from producers of horticultural crops (Rynk et al., 1992).

One of the primary uses of compost in horticulture is as a growing medium for plants. Approximately 60 percent of all nursery and greenhouse plants currently marketed are grown in containers. Because 60 to 70 percent of the container-growing medium is organic matter, the potential market for high-quality compost is substantial (Gouin, 1991). As with farmers, however, the high value of the horticultural industry's crops also make this sector very cautious and resistant to change (Alexander, 1990; Gouin, 1989). In addition, the horticultural industry already has a dependable supply of products containing organic material. One of these products is peat moss. Significant amounts of peat moss are used by nurseries for potting mixes. Compost could be used as a substitute for peat moss because it is a relatively inexpensive local source of organic matter. In order for compost to take over a substantial amount of the market share currently held by peat moss, laboratory analyses and field tests must be conducted to demonstrate the benefits, safety, and reliability of the material (see Chapter 9).

Silviculture

Silviculture or forestry applications are a potentially large market for compost. A national study estimated that the aggregate potential for silviculture application was 50 million metric tons annually (Slivka, 1992). Four segments of this market present viable opportunities: forest regeneration, nurseries, Christmas tree production, and established forest stands.

Regenerating forests represents the largest potential market for compost in a silvicultural application (Shiralipour et al., 1992). Results from limited experimentation with

Table 8-3. (Continued).

Potential Compost Users	Quality Requirements	
Agricultural Industry	High: Low concentration of physical/chemical contaminants[a] High organic content ½" particle size	No phytotoxicity Low soluble salts Good water-holding capacity
Landscaping Industry	High: Minimal odor pH 6.0–7.0; adjustable ≤½" particle size 2% moisture content	Low soluble salts No plant/human pathogens No weed seeds Dark color
Horticultural Industry/Nurseries	High: pH 6.0–7.0 ≤½" particle size Low soluble salts	Good nutrient content Low concentration of physical/chemical contaminants[a]
Public Agencies	High: Mature (stable) compost Low concentration of physical/chemical contaminants[a] ≤½" particle size Good nutrient content No weed seeds	Low: Able to support grass/wildflowers
Residential Sector	High: Minimal odor ≤½" particle size <40% moisture content	Low concentration of physical/chemical contaminants[a] Dark color
Other: Land reclamation Dedicated land Golf courses Sod farms	Low: Able to support grass Low Medium-high Medium-low	

[a]Physical contaminants are visible, noncompostable particles; chemical contaminants include heavy metals and toxic substances.
Sources: EPA, 1993; Rynk et al., 1992.

compost applications during forest regeneration have shown that compost applications have improved the physiochemical properties of soil and have led to excellent seedling survival and sustained growth advantages (Shiralipour et al., 1992). One long-term study, in which MSW compost was applied during forest planting, determined that MSW compost can provide forest growth advantages while causing no detectable problems (Shiralipour et al., 1992).

Forest nurseries and Christmas tree production represent potentially low-volume/high-value applications of compost. Organic amendments increase plant vigor, facilitate improved root proliferation, and enhance survival in outplanting (Shiralipour et al., 1992). Approximately 123.5 acres of forest nurseries in Florida produced approximately 106 million seedlings for a 1990 planting of 150,670 acres of new plantations (Shiralipour et al. 1992). An average of 53.5 tons per acre of organic matter are added annually to maintain productivity of the seed beds. Compost could be used in such applications (Shiralipour et al., 1992).

The option to use compost in established forests is not as attractive as those opportunities outlined above due to difficulties associated with gaining adequate access to these areas with compost spreading machinery. Recently planted forests, however, could be treated before canopy closure and while access still is possible (Shiralipour et al., 1992).

Public Agencies

Compost uses that are applicable to the public sector include land upgrade, parks and redevelopment, weed abatement on public lands, roadway maintenance, and median strip landscaping. Municipalities that produce compost should examine their internal needs for soil amendments, fertilizers, topsoil, and other products. Since many communities have this built-in market for compost, they can avoid spending funds on such products, adding to the overall cost-effectiveness of implementing a composting program. Some states have established standards (e.g., Florida, Iowa, Maine, Minnesota, New Hampshire, New York, and North Carolina) and/or procurement

> ### The Use of Yard Trimmings Compost in the Horticultural Industry
>
> Many facilities for the composting of yard trimmings successfully market their compost to the horticulture industry. Some municipalities have designed innovative marketing arrangements that benefit both the community and the user. For example, in Scarsdale, New York, the city works with a local nursery in the composting of approximately 35,000 cubic yards of yard trimmings per year and in the distribution of the final compost. In return for a share of the product, the nursery assists with turning the windrows and provides storage space for the finished compost. Twice a year, the compost is available free of charge to residents in a "giveaway" program. The remaining compost is marketed by the nursery as mulch and also blended into potting soil and topsoil.
>
> A composting facility located in Carver County, Minnesota, has set up an enterprising arrangement with the University of Minnesota. The composting facility is located at the university's landscape arboretum. In exchange for the site, the arboretum receives approximately one-half of the finished compost product.

preferences (e.g., California, Florida, Illinois, Iowa, Kentucky, Maine, Maryland, Minnesota, Missouri, Nebraska, New Jersey, North Carolina, Pennsylvania, South Carolina, Washington, and West Virginia) for using compost in public land maintenance activities funded by the state (Kashmanian, 1992) (see Chapter 7).

Public agencies can use both high- and low-quality composts. High-quality composts should be used in locations, such as parks and playing fields, where people or animals come in direct contact with the material or in the upgrade of public lands. Upgraded land requires less water to irrigate, has an increased resale value, and has a higher quality of soil (EPA, 1993). In parks, high-quality compost can be used primarily to build and maintain turf. A coarse compost that has low water-retention capability can be applied to areas where weed control is necessary.

Lower quality compost can be used for purposes such as land reclamation, landfill cover, and, possibly, large highway projects (EPA, 1993). Lower quality compost can be used by public agencies (as well as private companies) to establish vegetative growth and restore or enhance the soil productivity of marginal lands. Uses of compost in land reclamation include restoring surface-mined areas, capping landfills, and maintaining road shoulders polluted with heavy metals and organic pollutants.

Reclamation of mine-spoil areas can be an excellent end-use option for large quantities of compost. Compost is valuable for these sites because of its high water-holding capacity. When using MSW compost in mine-spoil reclamation, soil-plant ecology must be considered in regard to intended land use. For example, if the land is reclaimed for a natural area, the compost will be required to aid in the reestablishment of a natural ecosystem (Shiralipour et al., 1992). If the land is reclaimed for future home sites, the compost should aid in the support of typical landscape plantings and should not contain any pathogens.

Compost with excessive levels of heavy metals can be used only for landfill cover. The Composting Council's 1992 survey reports that several communities across the nation are using compost in the final capping of landfills. Escambia County, Florida, has been composting mixed MSW since September 1991. From the outset, the county planned to use the compost product for daily and final landfill cover. The material is suitable for use as landfill cover after four weeks of composting.

Most road shoulders are already polluted with heavy metals and organic pollutants from motor vehicles (Shiralipour et al., 1992). Therefore, the use of mixed MSW compost would not substantially contribute to the deterioration of environmental quality and could reduce the bioavailability of existing contaminants (Chaney, 1991). The compost must be capable of supporting roadside growth with minimal erosion, and the compost must comply with both state and federal standards for land application. Federal and state highway departments have standards or guidelines for reseeding and landscaping of highway shoulders that might need to be modified to enable use of compost. The growth of this end use depends on the amount of road construction and maintenance.

Residential Sector

The residential sector represents a substantial market for compost. Gardeners frequently use compost as a soil amendment to improve the organic matter and nutrient content of

> ### Municipalities Utilizing Compost for Public Works Projects
>
> Mount Lebanon, in Allegheny County, Pennsylvania, uses compost in parks and on the city's golf course (EPA, 1993). The compost is made from leaves collected in the community. The county also is planning to set up a series of composting areas in city parks and to make the finished compost available to municipalities and park departments. Compost produced in Hennepin County, Minnesota, is used by the county's parks department or redistributed to municipalities, which make the compost available to residents in bulk form free of charge (EPA, 1993). The compost is made from yard trimmings collected from residents and landscapers.

Table 8-4. Examples of compost quality guidelines based on end use.*

Characteristic	End use of compost			
	Potting grade	Potting media amendment grade [a]	Top dressing grade	Soil amendment grade [a]
Recommended uses	As a growing medium without additional blending	For formulating growing media for potted crops with a pH below 7.2	Primarily for top-dressing turf	Improvement of agricultural soils, restoration of disturbed soils, establishment and maintenance of landscape plantings with pH requirements below 7.2
Color	Dark brown to black	Dark brown to black	Dark brown to black	Dark brown to black
Odor	Should have good, earthy odor	Should have no objectionable odor	Should have no objectionable odor	Should have no objectionable odor
Particle size	Less than 1/2 inch (13 millimeters)	Less than 1/2 inch (13 millimeters)	Less than 1/4 inch (7 millimeters)	Less than 1/2 inch (13 millimeters)
pH	5.0–7.6	Range should be identified	Range should be identified	Range should be identified
Soluble salt concentration (mmhos per centimeter)	Less than 2.5	Less than 6	Less than 5	Less than 20
Foreign materials	Should not contain more than 1% by dry weight of combined glass, plastic, and other foreign particles 1/8–1/2 inch (3–13 centimeters)	Should not contain more than 1% by dry weight of combined glass, plastic, and other foreign particles 1/8–1/2 inch (3–13 centimeters)	Should not contain more than 1% by dry weight of combined glass, plastic, and other foreign particles 1/8–1/2 inch (3–13 centimeters)	Should not contain more than 5% by dry weight of combined glass, plastic, and other foreign particles
Heavy metals	Should not exceed EPA standards for unrestricted use [c]	Should not exceed EPA standards for unrestricted use [c]	Should not exceed EPA standards for unrestricted use [c]	Should not exceed EPA standards for unrestricted use
Respiration rate (milligrams per kilogram per hour) [b]	Less than 200	Less than 200	Less than 200	Less than 400

[a] For crops requiring a pH of 6.5 or greater, use lime-fortified product. Lime-fortified soil amendment grade should have a soluble salt concentration less than 30 mmhos per centimeter.

[b] Respiration rate is measured by the rate of oxygen consumed. It is an indication of compost stability.

[c] These are EPA 40 CFR Part 503 standards for sewage biosolids compost. Although they are not applicable to MSW compost, they can be used as a benchmark.

* These suggested guidelines have received support from producers of horticultural crops.

Source: Rynk et al., 1992.

the soil and to increase the soil's moisture-holding capacity. Compost also can be used as a top dressing and as a mulch. The amount of compost used by the residential sector depends on the ability of suppliers to consistently produce a quality product at a reasonable cost. For example, only high-quality compost with low soluble salt concentrations should be used for home gardens (Rynk et al., 1992). Such compost should have a good earthy color and odor and be free of clods. See Table 8-3 for a list of quality requirements for the residential sector.

Homeowners are becoming increasingly familiar with the composting of yard trimmings through community yard trimmings collection programs and promotional backyard composting campaigns. This familiarity encourages acceptance of yard trimmings compost as a high-quality product. Developing residential markets for mixed MSW-derived compost, however, might prove more difficult due to the reluctance of the residential sector to accept mixed MSW compost as a high-quality product.

In addition to product quality, other factors that affect the quantities of compost used by the residential sector include population growth, the economy, and the housing industry. Communities that have a large percentage of single-family homes typically have a higher demand for soil amendments than areas of high-density housing (EPA, 1993).

Summary

Compost provides a stabilized form of organic matter that improves the physical, chemical, and biological properties of soils. It is currently used by a wide range of end users, including commercial industries (e.g., agriculture, landscaping, horticulture, and silviculture), public agencies, and private citizens. There is great potential for expanding these end-use markets. To market compost successfully, yard trimmings and MSW composting facilities must learn the specific requirements of potential end-users for quality, composition, appearance, availability, and price of the product.

Using Compost as a Growing Medium

Researchers at the University of Florida conducted an experiment using mixed MSW-derived composts as growing medium for plants. The plants used in the experiment—the Cuban royal palm, orange jessamine, and dwarf oleander—are grown commercially in tropical and subtropical climates, primarily for landscaping. The study found that the growth rates of the palm and jessamine grown in mixed MSW compost were not significantly different than those grown in the potting mix used as a control medium; the oleander performed better in mixed MSW composts than in the control soil (see table below). The study concluded that the mixed MSW composts were no better or worse in terms of plant growth than the commercial potting mix, which is sold for $35 per cubic yard.

Growth of Three Tropical Landscape Crops as Influenced by MSW-Growing Media

		Av. height, cm			Av. dry weight at end of production, g.[1]		
Species	Medium	At planting	At end of production	Δ	Shoot	Root	Total
Cuban royal palm	Control	126.0	171.0	45.0 a	276.7	197.2	473.9 a
	Hydrolite	119.8	169.8	50.0 a	252.0	93.4	345.5 a
	Geophile	119.4	181.2	61.8 a	375.3	340.4	715.3 a
	Glatco-Lite	117.2	172.8	55.6 a	Samples destroyed in laboratory fire		
Orange-jessamine	Control	19.6	60.2	40.6 a	89.8	93.8	183.6 a
	Hydrolite	25.0	73.4	48.4 a	99.2	119.3	218.5 a
	Geophile	23.6	64.4	40.8 a	121.6	69.0	190.5 a
	Glatco-Lite	24.8	72.2	47.4 a	105.3	88.6	193.9 a
Dwarf oleander	Control	26.6	50.0	23.4 ab	140.7	49.4	191.1 a
	Hydrolite	28.0	47.4	19.4 a	143.8	84.0	227.7 ab
	Geophile	27.0	55.2	28.2 bc	207.1	93.8	300.9 bc
	Glatco-Lite	26.4	56.2	29.4 c	232.6	98.5	331.1 c

[1] Values are the averages of 5 determinations for the height measurements, 4 determinations for the biomass determinations. Within species, mean separation is by the least significant difference test at the 5% level.

Production time = 6 months for orange-jessamine and dwarf oleander, 12 months for Cuban royal palm.

Glatco-Lite is a compost produced from pulp and paper mill by-products; Geophile is a compost produced from mixed MSW that has undergone ferrous metals separation prior to composting; Hydrolite is produced from composting dewatered biosolids and MSW.

Source: Fitzpatrick, 1989.

Chapter Eight Resources

Alexander, R. 1991. Sludge compost use on athletic fields. BioCycle. July, 32(7):69-70.

Cal Recovery Systems (CRS). 1989. Composting technologies, costs, programs, and markets. Richmond, VA: U.S. Congress, Office of Technology Assessment. As cited in: U.S. Congress, Office of Technology Assessment. 1989. Facing America's trash: What next for municipal solid waste? OTA-0-424. Washington, DC: U.S. Government Printing Office.

Cal Recovery Systems (CRS). 1988. Portland area compost products market study. Portland, OR: Metropolitan Service District.

Chaney, R.L. 1991. Land application of composted municipal solid waste: Public health, safety, and environmental issues. p.61-83. As cited in: Shiralipour et al., 1992. Uses and benefits of municipal solid waste compost. Biomass & Bioenergy. Tarrytown, NY: Pergamon Press. 3(3-4):267-279.

Composting Council (CC). 1992. Quarterly Newsletter. October. Washington, DC: Composting Council.

Gouin, F.R. 1989. Compost standards for horticultural industries. BioCycle. August, 30(8):42-48.

Gouin, F.R. 1991. The need for compost quality standards. BioCycle. August, 32(8):44-47.

Hoitink, H.A.J., and P.C. Fahy. 1986. Basis for the Control of Plant Pathogens with Compost. Annual Review of Phytopathology. Vol. 24:93-114.

Hoitink, H.A.J., Y. Inbar, and M.J. Boehm. 1991. Status of compost-amended potting mixes naturally suppressive to soilborne diseases of floricultural crops. Plant Disease. November, Vol. 75.

Kashmanian, R. 1992. Composting and Agricultural Converge. BioCycle. September, 33(9):38-40.

Kashmanian, R.M., H.C. Gregory, and S.A. Dressing. 1990. Where will all the compost go? BioCycle. October, 31(10):38-39,80-83.

Mays, D.A., and P.M. Giordano. 1989. Landspreading municipal waste compost. BioCycle. March, 30(5):37-39.

Mecozzi, M. 1989. Soil salvation. Wisconsin Natural Resources Magazine. PUBL-SW-093-89. Madison, WI: Department of Natural Resources, Bureau of Solid and Hazardous Waste Management.

Rynk, R., et al. 1992. On-farm composting handbook. Ithaca, NY: Cooperative Extension, Northeast Regional Agricultural Engineering Service.

Shiralipour A., D.B. McConnell, and W.H. Smith. 1992. Uses and benefits of municipal solid waste compost. Biomass & Bioenergy. Tarrytown, NY: Pergamon Press. 3(3-4):267-279.

Slivka, D.C. 1992. Compost: United States supply and demand potential. Biomass & Bioenergy. Tarrytown, NY: Pergamon Press. 3(3-4):281-299.

Spencer, R., and J. Glenn. 1991. Solid waste composting operations on the rise. BioCycle. November. 32(11):34-37.

U.S. Congress, Office of Technology Assessment. 1989. Facing America's trash: What next for municipal solid waste? OTA-0-424. Washington, DC: U.S. Government Printing Office.

U.S. Environmental Protection Agency (EPA). 1993. Markets for compost. EPA/530-SW-90-073b. Washington, DC: Office of Policy, Planning and Evaluation; Office of Solid Waste and Emergency Response.

Chapter Nine
Product Quality and Marketing

Marketing plays a critical role in any composting operation. It is important to identify end users for the compost product early in the planning stages of a compost facility, since a customer's requirements will have a significant impact on the design and operation of the facility. This chapter provides information about the quality of compost that can be expected from yard trimmings and MSW composting programs. It also discusses the importance of specifications and testing when marketing a compost product. In addition, this chapter examines the factors that must be considered when attempting to tap into identified markets, including market assessment, pricing, distribution, user education, and public education.

Product Quality

Consistent and predictable product quality is a key factor affecting the marketability of compost. Each compost user has different requirements for quality, however. These requirements must be understood and planned for when designing a composting system so that compost quality can be matched to a user's specific requirements. For example, certain end uses of compost (e.g., application to crops) require the production of a high-quality product that does not pose threats to plant growth or the food chain. Other uses of compost (e.g., landfill cover) have less rigorous requirements (Section 8 discusses various end uses of compost). Some of the key concerns about the potential risks of composted yard trimmings and MSW are discussed in this chapter. It should be noted, however, that although the potential risk associated with biosolids compost has been extensively studied, less is known about mixed MSW composts. More studies and field demonstrations are necessary to address research gaps concerning potential environmental and health effects of MSW-derived compost.

Yard Trimmings Compost Quality

Compost derived from yard trimmings contains fewer nutrients than that produced from biosolids, livestock manure, or MSW (Rynk et al., 1992); at the same time, it contains fewer hazardous compounds and other contaminants than compost derived from biosolids, manure, or MSW (see below). Nevertheless, concerns about the presence of heavy metals (e.g., lead, cadmium, zinc, copper, chromium, mercury, and nickel) and pesticides in finished yard trimmings compost could affect its marketability.

In general, the levels of heavy metals in MSW compost made from yard trimmings are well below those that cause adverse environmental and human health impacts (Roderique and Roderique, 1990). Table 9-1 shows data on heavy metal content in yard trimmings compost from two facilities. The content of heavy metals in the compost varied, but in all cases was below soil concentrations of trace elements considered toxic to plants, as well as the maximum levels established in Minnesota and New York for co-composted MSW and municipal sewage biosolids (Table 9-2).

Yard trimmings compost also might contain pesticide or herbicide residues as a result of lawn and tree spray application. High levels of these chemicals could result in a phytotoxic compost (a compost that inhibits or kills plant growth). Generally, however, pesticides tend to have a stronger attraction to roots and soil than to yard trimmings. In addition, pesticides and herbicides that are found in yard trimmings feedstock are usually broken down by microbes or sunlight within the first few days of composting (Roderique and Roderique, 1990). This is supported by several recent studies.

A 1990 study found low levels of four pesticides (captan, chlordane, lindane, and 2,4-D) in leaf compost; all levels were below U.S. Department of Agriculture tolerance levels for pesticides in food (Table 9-3). Low levels of pesticides also were found in yard trimmings compost in Portland, Oregon, in 1988 and 1989 (Hegberg et al.,

Product Quality and Marketing

MSW Compost Quality

In order to market MSW compost successfully to many end users, concerns about potential threats to plants, livestock, wildlife, and humans must be addressed. One of the primary concerns is the presence of heavy metals (particularly lead) and toxic organic compounds in the MSW compost product. To date, where problems have occurred with MSW compost, they have resulted from immature composts, not metals and toxic organics (Chaney and Ryan, 1992; Walker and O'Donnell, 1991). Manganese deficiency in soil and boron phytotoxicity as a result of MSW compost application can be potential problems, however. Nevertheless, measures (including effective source separation) can be taken to prevent all of these problems and produce a high quality compost.

Heavy Metals and Organics

The bioavailability of contaminants in MSW compost describes the potential for accumulation of metals or organics in animals from ingested compost, or from food/feed materials grown on compost-amended soils. While research on the ingestion of MSW compost has only begun recently, field studies on biosolids and MSW

Quality Characteristics in Compost

Product quality depends upon the biological, chemical, and physical characteristics of the material. Some of the most desirable characteristics include:

- Maturity (e.g., properly cured and stabilized).
- High organic matter content.
- Absence of viable weed and crop seeds, pathogenic organisms, and contaminants (such as bits of glass, plastic, and metal).
- Proper pH for the designated end use (usually between 6.0 and 7.8).
- Available nutrients (e.g., nitrogen, phosphorus, and potassium).
- Low or undetectable levels of heavy metals and toxic organic compounds.
- Low concentrations of soluble salts (less than 25 mmhos [a measure of electrical conductivity]).
- Uniform particle size of less than 1/2 inch in diameter.
- Dark color and earthy odor.
- Moisture content below 50 percent.
- Absence of visual, noncompostable contaminants such as pieces of glass or plastic (Rynk et al., 1992; CRS, 1990).

It is important to note that storage practices can influence the quality of MSW and yard trimmings compost that is eventually marketed to end users. If piles of compost are not kept dry and aerated, anaerobic conditions prevail and odors and harmful anaerobic by-products will result (Rynk et al., 1992) (see Chapter 4).

Table 9-1. Heavy metals in yard trimmings compost.

Heavy Metals	Croton Point, New York	Montgomery County, Maryland[a]	Standard[b]
Cadmium (ppm)	ND	< 0.5	10
Nickel	10.1		200
Lead	31.7	102.7	250
Copper	19.1	35.5	1,000
Chromium	10.5	33.6	1,000
Zinc	81.6	153.3	2,500
Cobalt	4.2		NS
Manganese	374.0	1,100.0	NS
Beryllium	15.0		NS
Titanium (%)	0.09		NS
Sodium	1.51	0.02	NS
Ferrous	2.67	0.96	NS
Aluminum	3.38	0.66	NS

[a] Average of 11 samples, 1984-1985.

[b] For pesticides, standards are derived from USDA tolerance levels for pesticides in food (40 CFR Chapter 1, Part 180). For metals, standards are Class 1 Compost Criteria for mixed MSW compost, 6 NYCRR Part 60-5-3.

Source: Roderique and Roderique, 1990.

1991) (see Table 9-4). The chlordane concentrations in the Portland compost were believed to be a result of termite treatment around houses. Because chlordane is now banned from general use, the presence of this compound in compost should decrease in the future. The pentachlorophenol concentrations might be due to treatment of outdoor wood such as fenceposts. Preliminary studies conducted in Portland have shown that the presence of these compounds does not interfere with seed germination or plant growth (Hegberg et al., 1991).

As these studies indicate, levels of heavy metals and pesticide residues detected in yard trimmings compost have generally been insignificant. Nonetheless, composting facilities should test their product for these and other variables (including soluble salts, viable weed seeds, and pathogens), as described later in this chapter.

Table 9-2. Contaminant Limits for MSW compost (mg/kg).

Contaminant	Minnesota	New York
Cadmium	10	10
Chromium	1,000	1,000
Copper	500	1,000
Lead	500	250
Mercury	5	10
Nickel	100	200
PCB	1	1
Zinc	1,000	2,500

Source: Hegberg et al., 1991.

composts suggest that a small percentage of metals in compost-amended soil are bioavailable to plants and other organisms.

The bioavailability of lead in mixed MSW compost is of concern to some end users. Lead can present a potential risk to children who inadvertently ingest compost-amended soil. A study that examined the levels of heavy metals in MSW compost from five operating facilities found somewhat higher lead levels in MSW composts than the median level in biosolids (Walker and O'Donnell, 1991). Studies are necessary to determine if the bioavailability of this lead is reduced because of binding with hydrous iron oxide and phosphate in sewage biosolids compost. Based on available research, Chaney and Ryan (1992) conclude that lead concentrations in mixed MSW compost products should be limited to 300 mg/kg. MSW compost prepared from MSW separated at a central facility often contains lead concentrations of 200 to 500 mg/kg. Source separation of products containing lead should help reduce the concentration of lead in the compost product (see below). Diverting these materials from the MSW stream in the first place (through household hazardous waste collection programs) should further help reduce the level of lead in compost.

Researchers also have been concerned with food chain and dietary risks posed by another heavy metal, cadmium. As a result of research on cadmium risk conducted during recent years, it can be concluded that uncontaminated biosolids and mixed MSW composts pose no cadmium risk, even in extremely worst-case risk scenarios (Chaney and Ryan, 1992). Research also indicates that the bioavailability of cadmium is low, even in acidic soils. In general, absorption of heavy metals by plants increases if the soil is acidic (i.e., pH 7.0). In addition, because zinc (which is found along with cadmium in biosolids and MSW composts) interferes with cadmium uptake by plants, mixed MSW compost is even less likely to contribute cadmium to human and animal diets via plants (Chaney, 1991).

It should be noted that heavy metals also appear to become less soluble (therefore less bioavailable to plants) over time during full-scale mixed MSW composting. If the composting process is performed properly, metals become bound to humic compounds, phosphates, metal oxides, etc. in the compost and stay bound when mixed with soil (Chaney, 1991).

Toxic organic compounds, including polychlorinated biphenyls (PCBs), polycyclic aromatic hydrocarbons (PAHs), and polychlorinated aromatics (PCAs), are potential concerns with MSW compost. Research has shown that PCBs are quite stable in the presence of both natural soil bacteria and fungi (Nissen, 1981); therefore, any PCBs that do find their way to the feedstock will most likely be present in the compost. The concentration of PCBs in MSW compost has been found to be low, however. PAHs are another potential concern in MSW compost, degrading to acids that contribute to the phytotoxicity of unstable composts. PCAs also can pose some risk. While they have been found to bind to the organic fraction of compost, little information is available regarding their availability to organisms in the compost product (Gillett, 1992). More studies are needed to better determine the risks posed from toxic organic compounds in MSW compost.

Boron Phytoxicity

MSW compost contains substantial levels of soluble boron (B), which can be phytotoxic (Chaney and Ryan, 1992). Much of the soluble B found in MSW compost is from glues, such as those used to hold bags together (Volk, 1976).

Table 9-3. Pesticides in yard trimmings compost.

Heavy Metals	Croton Point, New York	Montgomery County, Maryland[a]	Standard[b]
Captan	0.0052		0.05-100
Total Chlordane	0.0932	<1.0[c]	0.03
Lindane	0.1810		1.00-7.00
Total 2,4-D	0.0025	<1.0	0.05-1.00

[a] Average of 11 samples, 1984-1985.

[b] For pesticides, standards are derived from USDA tolerance levels for pesticides in food (40 CFR Chapter 1, Part 180). For metals, standards are Class 1 Compost Criteria for mixed waste compost, 6 NYCRR Part 60-5-3.

[c] Average of 2 samples.

Source: Roderique and Roderique, 1990.

Table 9-4. Pesticide analysis of Portland, Oregon, yard trimmings compost.

Pesticide Classification	Residue	Number of Samples[a]	Samples Above Detection Limit[b]	Mean[c] (mg/kg)	Range[c] (mg/kg)
Chlorophenoxy herbicides	2,4-D	16	0	ND[d]	—
	2,4-DB	16	0	ND	—
	2,4,5-T	16	0	ND	—
	Silvex	16	0	ND	—
	MCPA	16	0	ND	—
	MCPP	16	0	ND	—
	Dichloroprop	14	0	ND	—
	Dicamba	16	0	ND	—
	Pentachlorphenol	14	9	0.229	0.001-0.53
Chlorinated Hydrocarbons	Chlordane	19	17	0.187	0.063-0.370
	DDE	14	3	0.011	0.005-0.019
	DDT	8	0	ND	—
	opDDT	14	2	0.005	0.004-0.006
	ppDDT	14	4	0.016	0.002-0.035
	Aldrin	16	1	0.007	0.007
	Endrin	16	0	ND	—
	Lindane	16	0	ND	—
Organophosphates	Malathion	14	0	ND	—
	Parathion	14	0	ND	—
	Diazinon	14	0	ND	—
	Dursban	15	1	0.039	0.039
Miscellaneous	Dieldrin	13	1	0.019	0.019
	Trifluralin	10	0[e]	—	—
	Dalapon	4	0	ND	—
	Dinoseb	5	1	0.129	0.129
	Casoron	8	0[e]	—	—
	PCBs	8	0	ND	—

a. The number of samples is the combined total for 2 sources of compost, which were sampled in June 1988, October 1988, April 1989, July 1989 and October 1989. The number of samples taken each time was not uniform (mostly 2 per period per source in 1988 and 1 per period per source in 1989).
b. The minimum detection limit is 0.001 ppm for pesticides and 0.01 ppm for PCBs.
c. Dry basis.
d. Not detectable (ND).
e. Residue detected but not measurable.

Source: Hegberg et al., 1991.

In general, B phytotoxicity has occurred when MSW compost was applied at a high rate to B-sensitive crops (e.g., beans, wheat, and chrysanthemums). It appears to be more severe when plants are deficient in nitrogen, when low humidity conditions are present, or when a great deal of transpiration occurs (e.g., as in greenhouses) (Chaney and Ryan, 1992). Because soluble B is more phytotoxic to acidic soils, liming can correct the problem. In addition, B phytotoxicity has been shown to be short lived; it seems to occur only in the first year of application (Chaney and Ryan, 1992).

Manganese Deficiency

Mixed MSW compost has been found to cause a lime-induced manganese (Mn) deficiency in soils in some cases (de Haan, 1981). Whether Mn deficiency will occur when mixed MSW compost is applied to soil depends on such factors as:

- *The pH of the soil* - Mixed MSW compost usually raises the pH of soil; when it is added to naturally low Mn acidic soils, the resultant high pH can cause Mn deficiency.

- *The susceptibility of the crop* - Crops that are susceptible to Mn deficiency include soybeans and wheat.

- *The clay content of the soil* - Mn concentration appears to increase with increasing clay content.

- *The height of the water table* - Soils that have been submerged during formation leach Mn and are more susceptible to Mn deficiency (Chaney and Ryan, 1992).

MSW compost producers need to consider the potential of mixed MSW compost to induce Mn deficiency, particularly if soils or crops in the area that the compost will be marketed are susceptible to Mn deficiency. If necessary, Mn can be added during composting to ensure that Mn deficiency does not occur.

The Effect of Source Separation

Many researchers support the use of source separation (see Chapter 3) to remove recyclable and nonrecyclable/noncompostable components from the compostable components. Source separation is key to reducing the heavy metal and visual contaminant concentrations in the finished compost. In a four-season discard characterization study in Cape May, New Jersey, at least 86 percent of metals found in MSW were attributable to noncompostable materials (plastic, wood, aluminum and tin cans, household batteries, etc.) (Rugg and Hanna, 1992). Another study examined the influence of preprocessing techniques on the heavy metal content in MSW compost. Looking at the level of heavy metals in finished compost and at different separation techniques, the study concluded that finished compost contained the lowest levels of zinc, lead, copper, chromium, nickel, and cadmium when source separation occurred (see Table 9-5). In practice, however, it might be very difficult to remove many of the materials containing heavy metals. Extensive separation once these materials have been mixed with organics can be very costly.

Product Specifications

Developing and utilizing appropriate compost product specifications ensures that high-quality compost will be produced. Specifications can be established for a number of parameters, including organic matter content, particle size, nutrient content (especially carbon-to-nitrogen ratio), presence of toxic compounds, nontoxic contaminant levels, concentration of weed seeds, seed germination and elongation, soluble salts, color, odor, and water-holding capacity (EPA, 1993).

All of these characteristics are critical to buyers. For example, high moisture content means customers receiving bagged compost receive bagged water as well. Particle size affects aeration, drainage, or water-holding capacity. The compost's pH, nutrient concentrations, or heavy metal concentrations restrict its usefulness for certain plants. If the compost is not stable, storage will be difficult and might affect the compost quality. Compost stability also has an impact on plant growth. Finally, presence of visible noncompostable contaminants might influence the buyer's perception of quality.

Table 9-5. Heavy metal concentrations in MSW-derived compost.

Metal	Processing method (mg/kg dry weight)			
	A	B	C	D
Zinc	1,700	800	520	230
Lead	800	700	420	160
Copper	600	270	100	50
Chromium	180	70	40	30
Nickel	110	35	25	10
Cadmium	7	2.5	1.8	1.0

A. **Mixed household wastes are composted without preparation.** The process takes approximately 12 months. After composting, the product is screened and inerts are removed.

B. **The collected household wastes are separated into two fractions.** The material contains most of the easily degradable organic material. Between two-and-a-half and five months are needed for this composting process.

C. **The collected wastes are shredded, then processed,** resulting in a fraction to be composted. This fraction is free of most inerts, such as glass and plastics.

D. **Wastes are separated at the source.** The organic components are collected separately at households. All necessary steps are taken to insure that components containing heavy metals do not enter the organic components.

Source: Oosthoek and Smit, 1987.

Uniform product specifications have not been developed for compost. A few states, however, have developed specifications and regulations for yard trimmings and MSW compost (see Chapter 7 for more information on legislation). During the planning stages of a composting facility, communities should determine what regulations and specifications, if any, have been established in their state. Specifications of bordering states also could be investigated in order to expand marketing options. Where a state has not established specifications, minimum acceptable product standards should be determined based on anticipated end uses.

The final compost product should exhibit the characteristics that are important to the customer. Prospective clients also can be provided with samples of the compost product and specification sheets listing the parameters tested and the results of the tests (a sample specification sheet is shown in Figure 9-1).

Product Testing

To ensure product quality, the compost product should be laboratory tested frequently. Many environmental laboratories test compost. A composite sample, composed of many small samples from different locations in the curing piles, will provide the most representative result.

Market Assessment

The best way to identify end users for a product is through a market assessment. The market assessment pinpoints potential consumers, along with their product requirements. Conducting this assessment in the early stages of the planning process and using the data as the basis for program design will increase the likelihood of widespread use of the final compost product and long-term stability of the composting program. In addition, a market assessment can estimate potential revenues from the sale of the compost. While the sale of compost is in general not a highly profitable activity, any revenues earned can help offset the cost of processing. Estimating revenues is also important in determining what equipment will be needed and what the facility's total budget will be. Figure 9-2 provides a sample market assessment form.

Once the market assessment is performed, potential users must be turned into real compost users. Many factors affect this transformation. The product must be priced,

Average Concentration of Essential Plant Nutrients (in percent)	
Kjeldahl Nitrogen (TKN)	1.40%
Phosphorus	1.56%
Potash	0.30%
Average Heavy Metal and PCB Concentrations (in microorganisms/g, dry weight basis)	
Cadmium	2.9
Copper	332.0
Lead	499.0
Mercury	4.6
Nickel	449.0
Zinc	929.0
PCBs	4.6

Source: Rohrbach, 1989.

Figure 9-1. Sample specifications sheet.

Among the tests most commonly conducted are those that determine the concentration of plant nutrients and toxic compounds. The compounds that are tested for will depend on the feedstock and any applicable regulations. Facility managers should be aware of possible heavy metal contamination in mixed MSW compost, or other contaminants introduced by specific sources. Some facilities also test for maturity and stability (by using growth germination tests and root lengths). The presence of weed seeds and phytotoxic compounds also should be monitored. Respiration rate determinations indicate the rate of decomposition to be expected; a reduction/oxidation test that measures aeration status of the compost can predict odor problems. Some composts actually suppress soilborne plant diseases, and that possibility should be assessed as well (see Chapter 8). The laboratory equipment requirements for tests of moisture content, pH, and particle size are minimal; an outside laboratory will be needed, however, to determine characteristics such as nutrient and heavy metal concentrations. Larger facilities perform respiration rate tests in house; smaller facilities will need to seek an outside laboratory. Finally, as an added selling point to potential users, field tests can be conducted, often by university staff or extension specialists at land grant schools, to demonstrate product utility and effectiveness.

It would be useful to carefully record the test data (on a computerized spreadsheet, if possible) to facilitate any reporting requirements that might have to be met and to provide a basis for comparing information collected over a long period of time. In this way, subtle changes in compost quality or properties can be observed.

A Successful Market Assessment for MSW Compost in Wright County, Minnesota

In Wright County, Minnesota, a product end-use market assessment was conducted as part of the county's plans to develop a state-of-the-art composting facility to manage a substantial portion of its MSW. Through the assessment, the county accomplished the following:

- Projected total county demand for compost products.

- Identified end-user specific requirements such as transportation, chemical and physical specifications, product pricing, application considerations, demand seasonability, and delivery schedules.

- Reviewed compost products' chemical and physical characteristics and related these to the various end uses and to applicable regulations.

- Identified market development activities such as field and laboratory testing tailored to local end-user requirements.

To identify end users, a questionnaire was mailed to over 130 potential users in a 15-mile radius of the proposed compost facility site. Data from local Chambers of Commerce were used to compile the list of potential end users. Twenty-two end users returned the questionnaire; these individuals were then personally interviewed (Selby et al., 1989).

Company Name _____

Contact Person _____

Address _____

Phone Number _____

Type of Business _____

1. **If you use or sell any of the materials listed below, please indicate the amount used or sold on an annual basis, as well as the cost per ton.**

Product Used	Amount Used (in tons)	Amount Sold (in tons)	Cost Per Ton
Composted manures			
Fresh manure			
Sewage sludge compost			
Mushroom compost			
Peat			
Loam			
Organic fertilizers			
Topsoil			
Potting soils			
Custom soil mixes			
Bark mulch			
Wood chips			

2. **At what percentage are your annual needs for the above items increasing or decreasing?**

3. **What are your current terms of purchase?**

Figure 9-2. Sample market assessment form.

4. If yard waste or MSW compost were available in quantity on an ongoing basis, how much would you purchase? Would the purchase terms differ from your current terms?

5. Under what conditions would you be prepared to negotiate a purchase agreement for compost?

6. What are your concerns when purchasing a compost product (for example, odor, price, nitrogen/phosphorus/potassium, fineness, packaging)?

7. When are your peak demands?

8. What are your transportation/delivery needs?

9. Would you be prepared to guarantee acceptance of a minimum quantity of compost?

Additional comments:

Please return to:
  J. Compost Farmer
  100 Dairy Road
  Poultryville, MA 00000

Adapted from: Rynk et al., 1992.

Figure 9-2. (Continued).

sold, and distributed, and the buyers must be educated so they can optimize their sales efficiency.

Private vs. Community Marketing

Communities can market compost themselves or rely on private companies that are in the business of marketing compost. Private marketers can advertise the product by attending trade shows, field demonstration days, and other events; developing a good public relations campaign; suggesting appropriate equipment for handling the compost; and pricing the compost competitively.

Municipalities also can perform all of these functions, but this might put a burden on available resources. Some communities find that the revenues received from marketing compost can offset administrative and promotional costs. Others find, however, that they do not have the in-house marketing expertise or a suitable infrastructure to administer a program and thus choose to enlist the services of a professional marketing company.

Communities that opt to market the compost themselves should check whether they have the legislative authority to market compost products. Cities with their own programs also enter into the sensitive area of competing for business in the private sector. Municipal employees who sell compost to markets such as chain stores and nurseries can be at a disadvantage compared to salespeople who work for private firms, especially in terms of flexibility in dealing with potential customers.

Another approach to marketing compost that is becoming increasingly popular is to market the product through a broker (CRS, 1990). The broker buys the compost at a low price and takes responsibility for product testing, compliance with regulatory constraints, and promotion. A compost broker in the Northeast buys compost from a number of municipalities in the region and resells it to a network of landscapers and major topsoil users.

Free Compost

Some facilities build a customer base by giving away compost. Middlebush Compost, Inc., has been composting leaves in Franklin Township, New Jersey, since early 1987. At first, in order to establish markets, the company gave the product away as part of its marketing campaign. By the end of 1989, they were able to sell all of their product at $10 per cubic yard screened and $6 per cubic yard unscreened. The compost was sold to landscapers, developers, nurseries, garden centers, and home owners for use as a potting soil, a soil amendment, or a mulch for water retention and weed control, and was also used to cap landfills (Meade, 1989).

Pricing

A number of factors play a role in determining the final price of the compost product, including compost quality and availability; the cost of the composting program; costs of transportation, production, marketing, and research and development; the price structure of competing products; and the volume of material purchased by an individual customer. Since the main objective of marketing is to sell the compost that has been produced, the price of the compost should be set to help achieve this. A logical strategy is to price the product modestly at first to establish it in the marketplace and then increase the price based on demand. Table 9-6 provides examples of prices established for yard trimmings and mixed MSW compost.

Several communities have not charged for compost in order to increase community awareness of the benefits of compost. Providing compost free of charge also promotes good will in a community and is an effective way to find commercial users who might be willing to try the

Table 9-6. Prices received for compost.

Facility or Community	Materials Composted	Market	Price
St. Cloud, Minnesota	Mixed MSW and Biosolids	Farm Fields, Landscapers	None or $4 per yard[a] $20 per ton
Portage, Wisconsin	Mixed MSW and Biosolids	City-Owned Industrial Park	$8 per yard[a]
New Castle County, Delaware	Mixed MSW and Biosolids	Landscapers, Horticulture	$4.50 per yard[a]
Sumter County, Florida	Mixed MSW	Nurseries, Sod Farms	Planning on $9-12 per yard[a]
Skamania, Washington	Mixed MSW	Homeowners	$5-10 per yard[a]
Montgomery County, Maryland	Yard Trimmings	Landscapers, Nurseries	$19.20 per ton[b]
Seattle, Washington	Yard Trimmings	Landscapers, Residents, City/County	$7.50-12.50 per yard[b]

Sources: [a]Goldstein and Spencer, 1990; [b]Taylor and Kashmanian, 1989.

product. Some experts warn against "giveaway" programs, however, because these can give the impression that the compost has no value. Many recommend charging at least $1 per cubic yard to associate value with the product.

Some communities charge a nominal fee to bulk users and nonresidents but give the product free to residents. Other communities charge residents a small fee. In Cleveland, Ohio, the Greater Cleveland Ecological Association, which serves 16 communities and composts approximately 250,000 cubic yards of leaves each year, sells compost to residents. Discounts might be given for large-volume buyers and for early payment. The pricing structure and whether to give the compost away are determinations that should be made on a community by community basis, depending on the amount of material available, its quality, and the opportunities for use (Mielke et al., 1989).

Location/Distribution Issues

Market location is of key importance for both product acceptance and transportation issues. Generally, the price of compost does not cover the cost of transportation over long distances (EPA, 1993). In most cases, therefore, the market for compost is within 25 or 50 miles of the composting facility (Rynk et al., 1992). Proximity to composting facilities promotes trust in the product through name recognition, increases buyers' access to the product, and enables the compost to be sold at a competitive price due to low transportation costs. Bagging the compost product can expand the potential market area. While bagging requires a higher capital investment in machinery and bags, the bagged product sells at a considerably higher price than most bulk compost. The higher price might justify higher transportation costs and, therefore, a larger market area (Rynk, et al., 1992). Municipalities are usually better off selling in bulk.

The cost of transporting compost also depends on its weight and bulkiness. Many compost products are marketed only locally because the bulkiness of the compost (400 to 600 kg/m^3 [700 to 1,000 lb/yd^3]) makes transportation expensive. Communities need to monitor available transportation funds carefully during facility planning stages so that the distance between potential markets and the manufacturing facility can be set accordingly.

Distribution systems for compost are diverse and often creative. A system should be developed based on a survey of the needs of the potential users. Most compost is distributed in the following ways:

- Direct retail sale or free distribution of bulk compost by truckload or in small quantities on site.
- Direct sale or free distribution of bagged compost on site or at special distribution centers.
- Direct sale or free distribution to wholesalers for processing in bulk or in bags to retailers (EPA, 1993).

Municipalities that perform composting should examine their own public sector markets and determine how much money is spent annually on fertilizer, top soil, and other soil amendments by governmental agencies in the region. A fair amount of demand often can be created internally by passing procurement ordinances specifying recycled materials. For example, bid proposals could require that the topsoil used for land reclamation contain a minimum level of compost.

Many facilities rely on local residents to transport the compost from the composting site. This approach is not always successful, as most residents can transport and use compost only in small quantities. Residential users also prefer bagged compost. Bagging requires additional investment in capital costs, which in turn requires higher pricing. A successful marketing program for bagged compost requires a high-quality product and intensive advertising to overcome price competition from competing products.

Cleveland's Options for Compost Distribution

The Greater Cleveland Ecological Association operates six facilities for composting yard trimmings and serves 16 communities in Cleveland, Ohio. The association sells compost in the following ways:

- Customers bring their own containers (bags or bushels) to the composting site; the cost is $0.75 per bushel.
- Customers pick up bulk loads of compost at the composting site. Customers' trucks are loaded for $13.50 per cubic yard.
- Compost is home delivered. There is a 2 cubic yard minimum, which is sold for $55.10, and a 10 cubic yard maximum, which is sold for $178.30. These prices include delivery and taxes. There is an additional charge of $20 for out-of-county delivery.
- Compost is bagged in 1 cubic yard plastic bags. These are sold through distributors who deal with the nursery and landscaping industries.

A discount is given for semi-truckloads delivered to landscapers and commercial growers to encourage the use of compost on lawns and in potting media for nursery stock. The compost has sold out every year.

Product Diversity

For both yard trimmings and MSW composting operations, producing a variety of products broadens the potential market base, increasing the amount of compost sold and the revenues earned. Producing more than one product can alleviate shortfalls during peak demand periods, thereby improving distribution and reducing the amount of storage space required. Producing several products from the feedstock also guards against generating an oversupply. Several grades of compost products (that would be significantly different in chemical or biological properties) could be manufactured by segregating portions of the feedstock. For example, a facility could offer soil amendment-grade and potting media-grade composts.

If such segregation is not possible, the compost produced at the facility could be modified to make several grades of compost. For example, the compost's nutrient properties could be supplemented, the pH adjusted to suit different market needs, or the particle size could be varied by using coarser or finer screens to manufacture a rough-grade and a fine-grade compost. Wholesalers and retailers of compost sometimes add other ingredients, like lime or sand, for special uses or markets. The costs of these options will vary according to region.

Education and Public Relations

The results of a marketing study carried out in Portland, Oregon, indicate that the quantity of compost used by residents is largely a function of public education and the capability of a facility to produce a high-quality product consistently (CRS, 1988). It is therefore important to work closely with potential end users to educate them about the product's benefits and how it should be used.

Product credibility as recognized by an independent third party could help improve sales. Communities can obtain several independent expert opinions to assure the user of the benefits of the product. These might include a representative of a university, an extension service, an agricultural experiment station, or even a large greenhouse, nursery, or farmer who has used the product and is willing vouch for it. Landscape or nursery associations might provide opportunities for composting facility representatives to speak at monthly meetings and offer educational information to their members. Once educational material is developed, the involvement of an educational network is vital (Tyler, 1992). The United States Department of Agriculture offers an educational program to farmers through the Cooperative Extension Service; communities can contact the Cooperative Extension Service to assist them in marketing finished compost to area farmers.

Throughout the marketing process, it is critical to present the compost as a usable product, not as a waste material that must be disposed of. It is imperative to realize the importance of a positive attitude and how contagious enthusiasm can be when presenting ideas on the uses of compost (Tyler, 1992). A positive approach can help reduce the potential stigma that users might assign to certain types of compost and promote acceptance of compost in the marketplace.

Updating the Market Assessment

Marketing requires continuous effort and does not stop once end users are secured. End users must become repeat customers if there is to be continued success of a marketing strategy. Monitoring of the marketplace is necessary to determine if all of the compost produced is being distributed, if users are satisfied with the product, and if the publicity strategies being employed are effective. Re-surveying potential users to determine whether they are now willing to use the compost is beneficial, as is updating the market assessment to identify any new markets that might have emerged since the last survey. Ongoing market surveys allow customers to participate in program development. Understanding customers' feelings and emotions paves the way for building trust in the compost product (Tyler, 1992).

Summary

The marketing of compost should be undertaken in the early stages of developing the composting facility in order to identify potential end users of compost and quality of compost they demand. This will assist decision-makers in all facets of planning, from designing the size of the facility to making financial projections of revenues. Communities should consider all aspects of marketing, including packaging, distribution, and applicable regulations. Marketing can be conducted in house or through marketing companies and brokers. Finally, community officials should keep in mind the constant need to gauge customer satisfaction and attitudes so that potential problems can be isolated and solved before they affect facility performance.

Chapter Nine Resources

Appelhof, M., and J. McNelly. 1988. Yard waste composting: Guidebook for Michigan communities. Lansing, MI: Michigan Department of Natural Resources.

Cal Recovery Systems (CRS) and M.M. Dillon Limited. 1990. Composting: A literature study. Ontario, Canada: Queen's Printer for Ontario.

Cal Recovery Systems (CRS). 1988. Portland-area compost products market study. Portland, OR: Metropolitan Service District.

Chaney, R.L., and J.A. Ryan. 1992. Heavy metals and toxic organic pollutants in MSW composts: Research results on phytoavailability, bioavailability, fate, etc. As cited in: H.A.J. Hoitink et al., eds. Proceedings of the International Composting Research Symposium. In press.

Chaney, R.L. 1991. Land application of composted municipal solid waste: Public health, safety, and environmental issues. As cited in: Proceedings of the Northeast Regional Solid Waste Composting Conference, June 24-25, 1991, Albany, NY. Washington, DC: Composting Council. pp. 34-43.

Composting Council (CC). 1992. Personal communication. Washington, DC.

de Haan, S. 1981. Results of municipal waste compost research over more than fifty years at the Institute for Soil Fertility at Haren/groningen, the Netherlands. Netherlands Journal of Agricultural Science. 29:49-61. As cited in Chaney and Ryan, 1992. Heavy metals and toxic organic pollutants in MSW composts: Research results on phytoavailability, bioavailability, fate, etc. As cited in: H.A.J. Hoitink et al., eds. Proceedings of the International Composting Research Symposium. In press.

Gillett, J.W. 1992. Issues in risk assessment of compost from municipal solid waste: Occupational health and safety, public health, and environmental concerns. Biomass & Bioenergy. Tarrytown, NY: Pergamon Press. 3(3-4):145-162.

Goldstein, N., and B. Spencer. 1990. Solid waste composting facilities. BioCycle. January, (31)1:36-39.

Harrison, E.B., and T.L. Richard. 1992. Municipal solid waste composting: policy and regulation. Biomass & Bioenergy. Tarrytown, NY: Pergamon Press. 3(3-4):127-141.

Hegberg, B.A., W.H. Hallenbeck, G.R. Brenniman, and R.A. Wadden. 1991. Setting standards for yard waste compost. BioCycle. February, 32(2):58-61.

Meade, K. 1989. Waiting for the leaves to fall. Waste Alternatives. March: 34-38.

Mielke G., A. Bonini, D. Havenar, and M. McCann. 1989. Management strategies for landscape waste. Springfield, IL: Illinois Department of Energy and Natural Resources, Office of Solid Waste and Renewable Resources.

New York Legislative Committee on Solid Waste Management. 1992. Earth for sale: Policy issues in municipal solid waste composting. Albany, NY.

Nissen, T.V. 1981. Stability of PCB in soil. As cited in: M.R. Overcash, ed. Decomposition of toxic and nontoxic organic compounds in soil. Ann Arbor, MI: Ann Arbor Science Publishers, pp. 79-87.

Oosthnoek, J., and J.P.N. Smit. 1987. Future of composting in the Netherlands. BioCycle. July, 28(7):37-39.

Pahren, H.R. 1987. Microorganisms in municipal solid waste and public health implications. Critical reviews in environmental control. Vol. 17(3).

Portland Metropolitan Service District (PMSD). 1989. Yard debris compost handbook. Portland, OR: PMSD.

Roderique, J.O., and D.S. Roderique. 1990. The environmental impacts of yard waste composting. Falls Church, VA: Gershman, Brickner & Bratton, Inc.

Rohrbach, J. 1989. Delaware Solid Waste Authority. New Castle, DE.

Rugg, M., and N.K. Hanna. 1992. Metals concentrations in compostable and noncompostable components of municipal solid waste in Cape May County, New Jersey. Paper presented at the Second U.S. Conference on Municipal Solid Waste Management, Arlington, VA.

Rynk, R., et al. 1992. On-farm composting handbook. Ithaca, NY: Cooperative Extension, Northeast Regional Agricultural Engineering Service.

Selby, M., J. Carruth, and B. Golob. 1989. End use markets for MSW compost. BioCycle. November, (30)11:56-58.

Smit, J.P.N. 1987. Legislation for compost in the Netherlands—Part II. As cited in: de Bertoldi, M. et al., eds. Compost: Production, quality and use. New York, NY: Elsevier Applied Science.

Taylor, A., and R. Kashmanian. 1989. Yard waste composting: A study of eight programs. EPA/530-SW-89-038. Washington, DC: Office of Solid Waste and Emergency Response, Office of Policy, Planning and Evaluation.

Tyler, R. 1992. Ground rules for marketing compost. BioCycle. July, 33(7):72-74.

U.S. Environmental Protection Agency (EPA). 1989. Charactcrization of Products Containing Lead and Cadmium in Municipal Solid Waste in the United States, 1970-2000. EPA/530-SW-89-015B. Washington, DC: Office of Solid Waste and Emergency Response.

U.S. Environmental Protection Agency (EPA). 1993. Markets for compost. EPA/530-SW-90-073b. Washington, DC: Office of Policy, Planning and Evaluation, Office of Solid Waste and Emergency Response.

Volk, V.V. 1976. Application of trash and garbage to agricultural lands. pp. 154-164. As cited in Chaney and Ryan, 1992. Heavy metals and toxic organic pollutants in MSW composts: Research results on phytoavailability, bioavailability, fate, etc. As cited in: H.A.J. Hoitink et al., eds. Proceedings of the International Composting Research Symposium. In press.

Walker, J.M., and M.J. O'Donnell. 1991. Comparative assessment of MSW compost characteristics. BioCycle. August, 32(8):65-69.

Williams, T.O., and E. Epstein. 1991. Are there markets for compost? Waste Age. April, 22(1):94-100.

Chapter Ten
Community Involvement

> Decisions involving solid waste management can generate considerable controversy among local residents, who are concerned about the safety and health of their friends and family as well as the welfare of their local environment. Therefore, it is crucial for local officials to develop strong support among their constituents whenever embarking on solid waste policy-making; planning a composting facility is no exception. Local support of a composting operation determines the ease with which the facility can be sited and the willingness of the public to participate in the program. Decision-makers can involve the public by inviting constituents to participate in many decisions surrounding composting facility siting, design, and operation. In addition, an educational and public relations program must be conducted to maintain citizen enthusiasm in the program once operation has commenced.

Planning the Composting Project

A well-run public information program can help generate support for a composting program in the planning stages. Local officials can undertake a publicity campaign to educate citizens and the media about how composting works and why it can be an effective waste management strategy. The publicity campaign can also point to the planned composting program as a source of civic pride because it is an indication that the community is environmentally aware. Publicity techniques can include:

- *Paid advertising* - Television or radio ads, newspaper ads or inserts, magazine ads, outdoor ads.
- *Public service advertising* - Radio announcements, free speech messages, community calendar notices, utility bill inserts.
- *Press coverage* - Briefings, news conferences, feature stories, press releases, press kits.
- *Non-media communications* - Presentations to civic organizations or schools, newsletters, exhibits/displays, mailings of key technical reports, promotional materials (brochures, door hangers, leaflets).

While promoting the benefits of composting, the public information program should also foster realistic expectations. Officials must provide honest and detailed information about issues such as:

- *Odor* - It is important to acknowledge that composting can generate odors, but that steps can be taken to minimize their impact on the surrounding community (see Chapter 6).

- *The portion of the waste stream that can be composted* - While as much as 30 to 60 percent of the MSW stream could potentially be composted, composting is not a panacea for managing all that a community discards.

- *Costs of composting* - The sale of compost will not generate enough revenue to support all the costs of a municipality's composting program. It is the cost savings from avoiding combustion or landfilling and the beneficial reuse of materials that make composting financially attractive.

Once the public has been informed about composting in general and about the proposed facility, the next step is to provide avenues through which members of the public can express their concerns. A variety of techniques are available for soliciting feedback from members of the public, including advisory groups/task forces, focus groups, telephone hotlines, public hearings, town meetings, referendums, interviews with people representing key groups or neighborhoods, and workshops to resolve specific issues. Some of these techniques give community members decision-making roles, while others give them advisory roles or simply provide information. The degree of decision-making authority might depend on what local laws require (some require that all policy and budgetary decisions be made by local officials) or what element of the composting program is under discussion and if it is controversial (for example, siting might require more decision-making by members of the community than details of processing technologies). Cornell University is currently doing a study on this subject, supported by funds from the Compost Council Research Foundation.

Community Involvement in Siting Decisions

Siting a composting facility can be a sensitive process for solid waste managers and site designers. The search for sites can be stalled by local residents who do not want a composting facility in their community (the *Not In My Backyard*, or NIMBY, syndrome). People might be especially opposed to siting a facility in populated areas or in areas located near residences, schools, and hospitals.

Residents near a site proposed for a composting facility might be concerned about potential problems with the operation, particularly about the potential for odor generation. Noise, traffic, visual impacts, and potential health threats might be additional concerns of residents. Officials should be prepared to listen to the public's concerns and to negotiate the site selection or the design of the facility. Many communities have changed site or facility design on the basis of citizens' concerns. Involving the public in siting decisions builds a greater sense of community solidarity in the project and facilitates compromise among the participants in the project.

Officials should assure residents that serious problems do not occur at properly managed facilities, and that effective corrective measures are available for any complications that do arise. However, it is important to communicate that composting is not risk free, just as combustion and landfilling are not risk free. Offering information about the experiences of other communities might help to allay concerns about the facility. Communicating information about any risks associated with the program is critical in building consensus for siting decisions. Because many misgivings among the public about solid waste management facilities are based on perceived risk, officials should be prepared to provide information dispelling or putting into perspective any fears that arise among community members. (Siting considerations and techniques for solving potential environmental problems are discussed in more detail in Chapters 5 and 6.)

Site selection committees should draft a set of objective criteria for choosing an appropriate location for the composting facility. A site selected on the basis of objective analysis of these criteria will be more acceptable to the public and will help counter any perceptions that the selection process is arbitrary.

Other guidelines for successful siting include:

- Accepting the public as a legitimate partner.
- Listening to the concerns of the different interests.
- Planning a siting process that permits full consideration of policy alternatives.
- Setting goals and objectives for public involvement and risk communication activities in each step of the siting process.
- Creating mechanisms for involving the public early in the decision-making process.
- Providing risk information that the public needs to make informed decisions.

Planning a Composting Facility Through Teamwork

Disposing of fish scraps from processing plants was a perennial problem for communities on the Maine shoreline until a consortium of public and private organizations found a solution through composting. The consortium included fish processors, the local water company, the U.S. Department of Agriculture, and the Maine Departments of Transportation, Environmental Protection, and Agriculture. By involving these groups in the planning process, adversarial relationships were minimized and a sense of joint ownership of the project developed.

One of the first steps of the consortium was to develop clearly stated goals. The groups also formed special action committees—management, budget and funding, public relations, and research—to carry out important tasks identified in the team planning sessions. The publicity committee, for example, was charged with addressing the concerns of local residents about composting. The committee developed a slide show for the public that depicted the process and benefits of large-scale composting, an educational project display for municipal offices, and a high-quality brochure. In addition, the publicity committee organized two public field days at the site and promoted the usefulness of finished compost through an information campaign. Today, due in part to this educational campaign, the facility continues to compost and has gained acceptance for its finished product. Many groups, such as the Maine Department of Transportation, along with towns and private citizens use the compost for landscaping and soil amendment purposes (York and Laber, 1988).

> ### Siting a Co-Composting Facility in Wisconsin
>
> Officials in Columbia County, Wisconsin, learned the importance of public participation while attempting to site a dual materials recovery facility (MRF) and in-vessel co-composting facility in 1989 and 1990. The officials selected a site for the facilities in Pacific Township. Meanwhile, the township board decided to exercise an ordinance granting it authority to approve the siting of any solid waste authority within township lines. When county officials purchased an option on the property prior to obtaining township approval and applied for a Wisconsin Department of Natural Resources permit, many township residents felt the county was trying to force the board to accept the site selection.
>
> County officials were confronted with a grave public perception and credibility problem. To avert further misunderstanding, the county notified residents in the area surrounding the proposed site and organized public hearings on the matter. At the same time, residents of the township established a citizens' committee to block the siting.
>
> Eventually, the county received a permit for the facility from the state Department of Natural Resources with the condition that the county agree to a number of clauses requested by local citizens. The county agreed, for example, to put a plastic membrane lining under the tipping floor of the MRF and to provide free collection of Pacific Township's garbage. Pacific Township also obtained authority to inspect the facility at any time during business hours and issue citations if anything was out of order. Thanks to the willingness of the county to listen to citizens' concerns and compromise on facility site and design, the MRF opened in the spring of 1991, and the co-composting facility opened in the fall of 1991 (ICMA and EPA, 1992).

- Being prepared to mitigate negative impacts on the community.
- Evaluating the effectiveness of public involvement and risk communication activities (EPA, 1990).

Public Participation in the Composting Project

To ensure that the composting project runs smoothly, members of the public must have a clear idea of their role in the program. Facility or community officials must communicate information such as the collection schedule, acceptable and unacceptable materials, and how the materials will get to the facility. Residents can be notified of collection dates by letter or through announcements in newspapers or on the radio.

Municipalities also can provide information to the public about home composting or leaving grass clippings on the lawn. This information can help reduce the amount of yard trimmings that a community needs to collect. For facilities that compost either yard trimmings or MSW, information also should be provided about the availability of finished compost and whether the product is free or for sale (Walsh et al., 1990).

At the composting facility, attractive and informative signs can communicate salient information to the public, including the nature of the project, the facility name, the hours of operation, and the business address and telephone number of the operator. Other signs can direct collection vehicles to unloading areas and indicate traffic circulation patterns. If there is a drop-off site, signs should guide people to the site and clearly present the rules for delivery of the materials. The facility operator should consider including a reception area in the plant and arranging for tours for interested members of the public and the media. Officials also can recruit volunteers from the community to participate in monitoring incoming materials and assisting at the drop-off facility.

> ### Educating Citizens About a Composting Program
>
> Information that can be provided to citizens in a notification about a composting program (and their role in the program) might include:
>
> - A statement of the intent and community benefits of a composting program.
> - A description of the intended uses of the compost.
> - A statement that compostable materials must not contain materials such as glass, metal, or household hazardous waste.
> - Instructions regarding the piling of yard trimmings, or if bags are used, the type of bag and bag closure to be used; for MSW composting, it will be essential to include information about source separation or commingling of compostables.
> - Instructions regarding the placement of the material at the curb or in the street.
> - The dates when materials will be collected in designated districts and the locations and hours of community collection stations and other drop-off locations.
> - A map showing designated drop-off collection areas (UConn CES, 1989).

Another way to maintain a positive relationship with the community is to establish a complaint response procedure. Some municipalities, for example, recruit residents to participate on "odor panels" that report to the facility the detection of any odors originating from the composting site. (Most composting facilities receive some complaints, primarily about odors.) Complaints should be logged, along with the time, name of the complainant, action taken in response to the complaint, and the date of followup communication to the complainant. This procedure is designed to ensure that small problems are solved before they become larger ones, and will reassure neighbors that their concerns are taken seriously (Walsh et al., 1990).

Community education about composting should continue after the composting operation begins to ensure that support does not wane. Ongoing publicity can describe successes in the composting project and remind the community that composting is an important tool to manage organic materials. The effectiveness of the publicity techniques should be evaluated periodically.

Community Education at the Marketing Phase

Community education also is important in marketing, especially if the compost will be distributed to residents. Literature can be developed explaining the merits and uses of the compost, how the compost will be distributed (e.g., in bags or in bulk), and any restrictions on use. Samples of the product also can be provided to potential users. In addition, some communities give away compost to residents or neighbors of the composting facility (or provide it at a nominal charge). This can be promoted as a public service. Such programs foster goodwill and build support for the composting facility, although communities should be aware that giving compost away could create the impression that it has no monetary value. Many giveaway programs require residents to pick up compost at a centrally located site, which is sometimes combined with a recycling center. This approach helps to raise public awareness about composting and recycling, and provides a tangible reward to residents for their efforts. If the compost will be distributed or sold to users other than residents, marketing research should be conducted and sales strategies devised (see Chapter 8).

Chapter Ten Resources

International City Management Association (ICMA) and U.S. Environmental Protection Agency (EPA). 1992.

> ## Summary
>
> Successful management of community relations requires the same degree of attention, systematic planning, and expertise as do the more technical elements of designing and operating a composting program. Officials should strive to involve the public directly in planning and siting the facility. Community involvement in decision-making builds a sense of ownership among local residents and minimizes confrontation among concerned parties.
>
> A strong educational program should accompany each phase of facility planning and operation. Public outreach should include risk communication, information on program logistics such as collection times and places, and public service announcements and advertisements aimed at raising public participation in the program.

Case studies of municipal solid waste facility sitings: Success in your community (revised draft report). 73-80.

Logsdon, G. 1991. Slowing the flow to the landfill. BioCycle. May, 32(5):74-75.

Richard, T.L., N.M. Dickson, and S.J. Rowland. 1990. Yard waste management: A planning guide for New York State. Cornell, NY: New York State Energy Research and Development Authority, Cornell Cooperative Extension, and New York State Department of Environmental Conservation.

U.S. Environmental Protection Agency (EPA). 1990. Sites for Our Solid Waste: A Guidebook for Effective Public Involvement. EPA/530-SW-90-019. Washington, DC: Office of Solid Waste and Office of Policy, Planning and Evaluation.

University of Connecticut Cooperative Extension Service (UConn CES). 1989. Leaf composting: A guide for municipalities. Hartford, CT: State of Connecticut Department of Environmental Protection, Local Assistance and Program Coordination Unit, Recycling Program.

Walsh, P.A., A.S. Razvi, and P.R. O'Leary. 1990. Operating a successful compost facility. Waste Age. March, 21(3):137-144.

York, C.E., and D. Laber. 1988. Two c's overcome NIMBY. BioCycle. October, 29(10):60-61.

Chapter Eleven
Economics

Sound financial planning is a crucial step in the successful development of a composting program. When considering the multitude of options available for tailoring a composting program to the needs and resources of the community, decision-makers must weigh the costs and benefits involved and determine whether composting represents a feasible management option for their community. This section describes economic factors that a community will need to examine when designing a composting program. To give decision-makers a clear framework of the costs and benefits involved in setting up and managing a composting facility, the primary assumption used throughout this chapter is that the community owns and operates the facility. Communities might want to examine other options, such as forming partnerships with other municipalities or private companies, hiring a contractor to run the facility, or trying to attract a private company to establish the facility (see Chapter 1 for more information on planning). A financial worksheet also is included at the back of the Chapter that can be used to analyze cost information (Figure 11-1).

Cost/Benefit Analysis

Costs for developing a composting program typically include 1) capital costs for establishing and equipping a facility and 2) operation and maintenance (O&M) costs associated with such activities as collection, transportation, processing, program administration, and marketing. Communities also must keep in mind the revenue-generating or cost-avoiding aspects of the various composting choices. Composting can offer several potential economic benefits to communities:

- Extended landfill longevity.
- Avoided costs from reducing or eliminating the need for soil amendment purchases.
- Reduced or avoided landfill or combustor tipping fees.
- Environmental benefits from reduced landfill and combustion use.
- Creation of new jobs.
- Revenues from selling the finished product.
- Revenues from sale of recyclables.

The net cost of a composting program can be projected by estimating all capital and O&M costs and subtracting any revenue and/or avoided costs generated from running the program. This type of economic assessment, called a cost/benefit analysis, is used widely throughout the government and private industry to determine the cost-effectiveness of implementing a social program or making an investment. To be effective, cost/benefit analyses should be as comprehensive and detailed as possible. Many communities, therefore, hire consultants to conduct this analysis.

Decision-makers should not expect to earn money from composting. Most community owned and operated management facilities function at some expense to the taxpayers in the area. This should not diminish the feasibility of instituting a composting facility, however. Instead, decision-makers should compare the costs of composting against the costs of landfilling and combustion. With the rising costs of landfilling and combusting, composting programs frequently prove to be economically sensible management options.

Communities can choose from a host of collection methods, site designs, and equipment technologies when planning a composting program. For instance, implementing a simple composting program for yard trimmings that requires residents to drop off their materials would require minimal capital and operating expenses from the community. In contrast, MSW composting programs typically entail far greater start-up and operating expenditures and are often constructed to serve more than one community. Typically, the program design that a community selects for a composting project depends on the desired level of capital expenditure and on resources, such as equipment and labor, that are already available and can be partially or wholly allocated to the composting program.

Capital Costs

Capital expenses must be evaluated when establishing a compost facility. First, the community must apply resources to plan the composting facility. This involves allocating resources to hire staff or consultants to design the facility, to hold community meetings, and to conduct outreach measures to communicate with the community on such issues as siting. Suitable land then must be located and purchased, the site must be prepared for the composting activity, and vehicles and equipment might need to be purchased. Some states might require composting facilities (particularly MSW facilities) to obtain an operating permit, a process that can involve considerable assistance from staff and/or consultants. When projecting yearly costs of a composting operation, communities should annualize capital expenses for equipment and site preparation on the basis of the depreciation rate and the discount rate.

Site Acquisition

The first capital expense that a municipality must consider is site acquisition. The cost of purchasing a site will depend on local real estate costs and on how centrally it is located. More remote sites likely will require less capital to obtain, but transportation costs will be higher. Communities that have land available should base the cost of using the site for a composting facility on the rental market value of the land.

Site Preparation/Land Improvements

Site preparation costs can vary widely, depending on the size of the planned facility and natural characteristics of the land. Communities will need to engage an engineer to design the site and the facility itself. Decision-makers must include in the economic analysis for the program the engineer's salary, even when assigning a staff engineer to design the composting facility.

Most sites for composting yard trimmings will require grading to give the processing area the ideal gradual slope to facilitate proper drainage and efficient composting. It might be necessary to construct drainage channels to improve control of any runoff. The state of Michigan estimates that these minimal preparation measures for a facility that composts yard trimmings will total about $17,000 on average for a small operation on a 4-acre site (Appelhof and McNelly, 1988). Significant variables in this estimate include the size of the site, the cost of labor, and the difficulty of grading the slope of the site. Infrastructure and construction costs are additional expenses to consider. Simple, seasonally oriented operations for the composting of yard trimmings are the least expensive to build, since minimal infrastructure is required. Road systems can be limited and unsurfaced, and fencing can be limited to the processing area to protect onsite equipment. For security, a gate on the access road can be constructed, as well as a simple gate house and office for onsite administration. Construction on this scale, for a medium-sized operation of 12 acres, has been estimated to cost $72,000 (Mielke, et al., 1989). These costs can vary widely, however. Paving the surface is the largest component in this estimate, but this cost might not be necessary, depending on the soil conditions at the site.

Larger facilities for composting yard trimmings will need to construct more road systems; to construct a fence around most of the perimeter of the property; and to maintain several buildings for equipment, maintenance, and administration. Many facilities opt to cover the compost pad to provide shelter from inclement weather. These operations will require higher capital expenditures. In addition, utility hookups will be needed. The main variable in this expense is the distance of the site from local services such as power lines and water mains. Finally, if a community chooses to implement drop-off collection in its composting program for yard trimmings, it also must consider the land needed for a drop-off area, including an area situated at the composting facility itself or areas located at several transfer stations where residents can deliver leaves, grass, and/or brush.

MSW facilities require significant site preparation to guard against runoff and leachate (see Chapter 6). These facilities will need to construct a drainage system to direct leachate away from the composting pad to a treatment area. In addition, a typical 300- to 400-ton per day MSW composting facility would require an office or administration area, a mixed processing building, and a composting area, which might or might not be fully enclosed. Typical MSW windrow composting facilities require 9 to 24 acres for the total facility. Capital outlays easily can exceed $1 million when preparing a site for MSW composting (Resource Systems, Inc. et al., 1990).

Vehicle and Equipment Procurement

Once the site has been prepared, communities must procure equipment. Again, lower technology operations for the composting of yard trimmings will have minimal start-up costs. Many small facilities that compost yard trimmings can operate with only a front-end loader for windrow turning; depending on the size and horsepower of the selected model, front-end loaders cost from $55,000 to $125,000 (UConn CES, 1989; Appelhof and McNelly, 1988). For higher throughput operations designed to accelerate the compost process, grinders or shredders for particle reduction are necessary; these cost approximately $40,000 to $90,000 (Wirth, 1989; UConn CES, 1989), depending on capacity. Screening equipment might be necessary for programs that seek to produce a high-quality compost; these units typically range from $25,000 for portable screens to $50,000 for stationary units (Appelhof and McNelly, 1988).

Municipalities planning small- or medium-sized operations for the composting of yard trimmings can share existing equipment with their public works department or other communities to reduce start-up costs. In addition, they might examine the possibilities of renting equipment. Large-scale facilities that compost yard trimmings might be interested in specialized windrow-turning equipment, which can process material more quickly than front-end loaders. While such equipment can increase efficiency, these units can cost well over $125,000.

MSW composting facilities, especially those with on-site MSW separation, typically require significant equipment purchases. Magnetic separators and vibratory screens, which are basic units commonly used in the separation process, can cost $5,000 and $20,000, respectively (Wirth, 1989). Shredders, grinders and trommels to process the feedstock each cost over $100,000. Input and output conveyors, which move feedstock to and from the different preprocessing equipment, vary in cost according to length, but can cost well over $100,000 for a 300 to 400-ton per day facility (Wirth, 1989). Other equipment that can be used for MSW composting includes odor control equipment (see Chapter 6), in-vessel windrow turning systems, and aeration equipment. Each of these systems are priced over $100,000 for the most simple versions of the technology. Equipment costs, advantages, and disadvantages are listed in Tables B-1 through B-8 in Appendix B.

Training

There are also start-up costs associated with personnel training. Whether a site is small and needs only a few part-time workers or has a large, onsite staff, training in equipment operations, administration, and, most importantly, quality control will be required. It is crucial that employees recognize the role they play in the production of consistent, highly marketable compost. Employee interest in the compost product begins with training, and proper training prevents extensive, costly trial-and-error learning periods (Appelhof and McNelly, 1988). (Chapter 6 contains more information on safety and health training.)

Permits

Communities must consider outlays associated with permitting. Permitting requirements vary from state to state, but usually a municipality seeking to open a composting facility must submit a comprehensive application detailing site design and operations. Permit applications typically include provide an engineering design report and a description of the site layout, facilities, and equipment. Information on specific site activities, such as active composting, monitoring, and product marketing, as well as a plan for preventing any environmental contamination of the site also should be included. Applications also might include personnel training information. Experts in the fields of engineering, compost science, finance, and law are usually needed to prepare applications. Planners should check with their state to determine the exact permit applications requirements.

Operating and Maintenance (O&M) Costs

O&M costs are those expenses that are incurred from running and managing a composting facility. Typical O&M costs include salaries, utilities, insurance, and equipment repair. These costs should be estimated during the planning process to determine the feasibility of the composting program for the community.

Collection Costs

One of the largest cost factors connected with any composting program is the type of collection system used. For a management system to be successful, the costs of collection must not exceed available resources. Solid waste managers should become familiar with all of the various options for separating and transporting materials to their management facility in order to select the method that will optimize their available resources (see Chapter 3). The O&M costs of a collection program vary according to the features of the collection method employed and certain variables unique to each community. These variables include local labor costs and the presence or lack of existing collection equipment and infrastructure.

Drop-Off Collection Costs for Yard Trimmings

Limited operating costs are associated with drop-off collection programs. Decision-makers must consider expenses for an ongoing education and communication program to encourage participation. Public officials can notify residents about the program via press releases and public service announcements. Informational pamphlets or brochures also can be mailed directly to residences, and public meetings can be held to discuss the program. Publicity campaigns can become expensive, however, since the process must be continuous in order to maintain the community's interest and participation (see Chapter 10 for more information on community outreach).

MSW Curbside Collection Costs for Yard Trimmings

Communities looking at curbside collection as a way to encourage greater participation must decide if such a program will be cost-effective by calculating the capital and O&M costs associated with the various types of curbside collection programs (bulk or containerized). Curbside collection is a more costly collection method than drop-off programs, but often the additional feedstock reduces unit processing costs.

Bulk collection systems for yard trimmings are a fairly labor-intensive undertaking. Personnel must spend considerable time per stop to collect the yard trimmings, resulting in higher operating costs than for containerized collections. This method also involves additional training expenses. Since bulk collections are more prone to contamination than containerized collections (particularly in communities in which tipping fees are charged to residents for their solid waste), collection personnel must be trained to spot and remove noncompostables hidden in curbside piles of yard trimmings. A curbside collection program that picks up containerized yard trimmings is a less labor-intensive operation. Such a program, however, does involve the purchase of the containers and their distribution to local residents. Chapter 3 contains detailed cost information on the various types of bags and bins. With some bins, collection trucks might require special lifting equipment.

MSW curbside collections can be conducted with source-separated or commingled MSW. Onsite separation will result in a large volume of recyclable and noncompostable material, and the latter must be transported off site for proper disposal. Source-separated MSW collection involves the costs of a continuing education program to inform residents on which components should be separated out. Commingled collection entails intensive sorting and removal prior to composting. Significant labor and capital expenditures will be incurred from installing and operating the needed preprocessing equipment. In addition, this collection procedure is not entirely free of added hauling costs.

Labor Costs

The labor required at a compost facility is contingent upon the volume and type of material handled, as well as the level of technology used. At a minimum, most operations require workers to receive and prepare compostable material for windrowing; form and turn the windrows; prepare the compost product for delivery; and perform monitoring, maintenance, and administration functions. A low-technology leaf composting site, one that processes about 3,000 to 5,000 cubic yards of leaves per acre with windrows turned by a front-end loader, could function with just two people working part time—one to operate the front-end loader and one to monitor the site and to water the windrows—or one full-time staff person. It has been estimated that such a facility would need about 135 to 150 labor hours to produce compost.

As the complexity of the facility and the program grows, more employees will be needed to perform various functions in the process. A high-technology site that composts yard trimmings and uses forced aeration and windrow turners to compost 80,000 cubic yards of feedstock per year, for instance, might need a plant manager or supervisor to oversee the site; equipment operators to handle the machinery and vehicles; and workers to empty bagged material, wet incoming compostable material, and maintain the site. Other workers could include a tipping floor operator, scale operator, and maintenance personnel. For a facility of this size, much of the staff would likely be employed full time.

Because of the amount of separation and preparation involved, mixed MSW composting facilities usually incur the greatest labor costs. In addition, at mixed MSW facilities, more extensive administration and maintenance is needed over all site operations. The compost process, in particular, must be overseen carefully and detailed records on each composting phase must be kept in order to ensure that a consistent product is produced. This labor drives up costs. For example, the Delaware Reclamation Project, a 1,000-ton per day mixed MSW composting site that sorts out noncompostable material with mechanical sorting and uses an in-vessel system for composting, requires an annual personnel budget of several million dollars.

Fuel, Parts, and Supplies

The O&M costs for facility equipment also can be significant. To operate as cost effectively as possible, fuel, oil, parts, and other supplies must be available to keep site machinery functioning at capacity. As a rule of thumb, municipalities can calculate these expenses for a yard trimmings facility as a percentage of the initial equipment capital costs, with estimates likely ranging around 15 percent. MSW composting will have higher equipment operating costs than yard trimmings facilities, since much of the composting is dependent on processing equipment.

A Public/Private Co-Composting Venture

In 1988, Gardener's Supply, a national mail order firm located in Burlington, Vermont, proposed to the city that it convince residents to drop off their leaves and lawn clippings at a 2-acre plot near the firm's headquarters. To supplement the yard trimmings, Gardener's Supply brought in 70 truckloads of cow and chicken manure and supervised the laying out of long windrows across the plot. A vigorous public education campaign consisting of flyers and signs combined with incentives from Gardener's Supply, such as coupons for free finished compost and discounts on the company's products, brought enough materials to create 500 tons of compost during the first year of operation. The public relations campaign cost about $2,400 (ICMA, 1992).

For example, at a 300- to 400-ton per day MSW in-vessel composting facility, these expenses could reach about $150,000 (Wirth, 1989). O&M costs for odor control equipment alone can range up to $360,000 annually, depending on the type of equipment used (see Table B-8 in Appendix B). These costs include those for biofilter media, chemical solutions for wet scrubbers, and carbon replacement for carbon absorption systems.

Outreach and Marketing Costs

The success of any composting program relies heavily on the individuals contributing the feedstock. The importance of public education in developing a composting program should not be underestimated; the composting program should be kept in the public's attention constantly in order for a community to maintain good participation and recovery levels. Education can take on a multitude of forms, from radio and television announcements to newspaper press releases. Communities should take advantage of as much "free press" as possible. Expenditures on public outreach often depend on the level of sophistication communities choose for their publications and other informational activities. A simple brochure or fact sheet can be written and printed for only a few cents per copy, for example. (Chapter 10 describes public outreach techniques, and Appendix A contains examples of public outreach material.)

Communities also can choose to market their finished compost to a variety of potential end users (see Chapters 8 and 9). Marketing efforts should commence with a market assessment to identify such factors as the transportation needs and desired chemical and physical specifications of each potential buyer. Municipalities often engage private companies to conduct these surveys and to develop creative advertising campaigns.

Other Costs

Lesser O&M costs, from utility payments to building and grounds maintenance, are inherent in any composting program and should be anticipated. Laboratory testing for monitoring the quality of the compost produced is another O&M cost. In addition, virtually all composting operations produce residual waste that must be disposed of. Large mixed MSW composting sites that receive commingled solid waste and sort out the noncompostable fraction will generate substantial volumes of reject material, often between 10 and 30 percent of incoming materials (Goldstein and Spencer, 1990). Yard trimmings facilities usually receive compostable yard trimmings separated from solid waste, and therefore extract a smaller percentage of residual waste, ranging from 1 to 10 percent (Kashmanian and Taylor, 1989). The specific costs of rejection disposal depend on the distance of the composting facility from the landfill, as well as on the tipping fees for the local landfill.

Benefits From Composting

Avoided Costs

The potential for avoided costs must be incorporated into the cost/benefit analysis of a composting facility. There are five major avoided costs associated with composting. First, because composting reduces the need for landfilling or combustion, some tipping fees are avoided. The amount of money saved through composting can be substantial, especially in communities where landfill or combustion capacity is scarce. In some areas, landfill or combustor tipping fees exceed $100 per ton. Second, a composting program extends current landfill life and delays the construction of a more expensive replacement landfill or incinerator. This is particularly significant for municipalities whose landfills are nearing capacity. Third, composting avoids the environmental costs of landfilling operations. For example, risks such as the production of leachate or methane gas are often not reflected by the tipping fees paid to dispose of solid waste; composting reduces these risks, although quantifying the amount of risk reduction might be a difficult task. Fourth, with composting the community saves money it currently spends on soil amendments, topsoil, mulch, wood chips, and other products for municipal landscaping, landfill cover, and reclamation programs. If a community uses the finished compost it produces for these purposes, it will avoid such expenditures. Finally, composting might result in costs that can be avoided through reduced trash collection. If drop-off or curbside programs divert enough yard trimmings or compostable MSW, sanitation personnel might spend less time collecting waste destined for the landfill or combustor.

Revenues

It is possible for communities to produce and market a high-quality product as a result of their composting efforts. These revenues can help defray some of the costs associated with a composting program; it is very unlikely, however, that these revenues alone will offset start-up and O&M costs. Compost from yard trimmings currently is more marketable, although markets for MSW compost might be opening up.

If revenue from the sale of compost is reported as the price per ton of finished compost, communities should calculate the ratio of tons of finished compost to tons of compost feedstock (e.g., $50/ton of finished compost where 5 tons of feedstock are used to produce 1 ton of finished compost would translate into $10/ton of feedstock revenue stream).

Limited additional revenues might be earned by separating out recyclable materials during the collection process or at a mixed MSW composting facility. Finally, if a community accepts yard trimmings or MSW for composting from neighboring communities, revenue can be generated by collecting tipping fees.

Summary

The cost components of the various composting systems are the major determinants in choosing a composting system. Judging whether a composting program will save money is difficult and depends as much on local circumstances as on the chosen combination of collections and processing. A municipality's size in proportion to its labor rates, land lease or purchase costs, and equipment cost and operating rates will determine much of its composting costs. While it is impossible to consider every contingency, planners must approach the issue of costs and benefits from this perspective, drawing all relevant factors into the equation to make a sound decision on composting in their community. To determine the savings and thus the economic feasibility of a composting facility, planners should evaluate the cost per ton of material composted and compare these numbers with the costs of alternative management options.

Chapter Eleven Resources

Appelhof, M., and J. McNelly. 1988. Yard waste composting guide. Lansing, MI: Michigan Department of Natural Resources.

Dickson, N., T. Richard, and S. Rowland. 1990. Yard waste management: A planning guide for New York State. Albany, NY: New York State Energy Research and Development Authority, Cornell Cooperative Extension, and New York State Department of Environmental Conservation.

Goldstein, R. and B. Spencer. 1990. Solid waste composting facilities. BioCycle. January, 31(1):36-39.

International City/County Management Association (ICMA). 1992. Composting: solutions for waste management. Washington, DC: ICMA.

Kashmanian, R., and A. Taylor. 1989. Costs of composting vs. landfilling yard waste. BioCycle. October, 30(10):60-63.

Massachusetts Department of Environmental Protection (MA DEP). 1991. Leaf and yard waste composting guidance document. Boston, MA: Division of Solid Waste Management.

Mielke, G., A. Bonini, D. Havenar, and M. McCann. 1989. Management strategies for landscape waste. Springfield, IL: Illinois Department of Energy and Natural Resources, Office of Solid Waste and Renewable Resources.

Resource Systems, Inc., Tellus Institute, and E&A Environmental Consultants. 1990. Lowell-Chelmsford Co-Composting Feasibility Study.

University of Connecticut Cooperative Extension Service (UConn CES). 1989. Leaf composting: A guide for municipalities. Hartford, CT: State of Connecticut Department of Environmental Protection, Local Assessment and Progress Coordination Unit, Recycling Program.

U.S. Environmental Protection Agency (EPA). 1993. Markets for compost. EPA/530-SW-90-073b. Washington, DC: Office of Policy, Planning and Evaluation, Office of Solid Waste and Emergency Response.

Wirth, R. 1989. Introduction to composting. St. Paul, MN: Minnesota Pollution Control Agency.

I. **START-UP (CAPITAL) COSTS**

  Site Preparation

   Engineering design _____

   Site clearing _____

   Grading _____

   Drainage _____

   Pad material _____

   Equipment

   Thermometers (2) _____

   (For other equipment, see optional costs)

                    TOTAL ONE-TIME START-UP COSTS: $ _____

                 TOTAL AMORTIZED START-UP COSTS/YR: $ _____

II. **OPERATIONAL COSTS**

  Labor

   Monitoring incoming materials
   and directing vehicles _____

   Forming windrows (loader operator) _____

   Turning windrows (loader operator) _____

   Watering windrows _____

   Monitoring temperature _____

  Fuel and Maintenance

   Front end loader _____

  Related Costs

   Lab analysis of compost _____

   Marketing/distribution compost _____

   Public education _____

   Other _____

                        TOTAL OPERATIONAL COSTS/YR: $ _____

Figure 11-1. Composting economics worksheet.

Economics

III. **OPTIONAL OPERATIONAL COSTS**

  <u>Equipment, Related Labor and O & M</u>

	AMORTIZED PRICE OF EQUIPMENT	LABOR	O & M	
Shredder	_____	_____	_____	
Screener	_____	_____	_____	
Chipper	_____	_____	_____	
Windrow turner	_____	_____	_____	
Other	_____	_____	_____	
Subtotal:	_____ +	_____ +	_____ =	_____

  <u>Other Optional Labor</u>

  Debagging _____

  Other _____

                                   Subtotal: _____

          TOTAL OPTIONAL OPERATIONAL COSTS/YR: $ _____

IV. **OPTIONAL COLLECTION COSTS**

  <u>Equipment, Related Labor and O & M</u>

	AMORTIZED PRICE OF EQUIPMENT	LABOR	O & M	
Compactor truck	_____	_____	_____	
Loader w/claw	_____	_____	_____	
Vacuum truck	_____	_____	_____	
Dump truck	_____	_____	_____	
Street sweeper	_____	_____	_____	
Other	_____	_____	_____	
Subtotal:	_____ +	_____ +	_____ =	_____

          TOTAL OPTIONAL COLLECTION COSTS/YR: $ _____

Figure 11-1. (Continued).

V. COST/BENEFIT ANALYSIS: COMPOSTING VS. CURRENT DISPOSAL

TOTAL COSTS

A. Total amortized start-up costs/yr $_____

B. Total operational costs/yr $_____

C. Total optional operational costs/yr $_____

D. Total optional collection costs/yr $_____

E. **Total Costs/Yr (A + B + C + D)** $_____

TOTAL BENEFITS

F. Avoided disposal cost/yr $_____

G. Avoided purchases of soil amendments/yr $_____

H. Projected income from sale of compost/yr $_____

I. **Total Benefits/Year (F + G + H)** $_____

TOTAL NET SAVINGS OR COST

J. <u>Net Savings/Year</u> (I - E if I > E) $_____

K. Net Cost/Year (E - I if E > I) $_____

Source: MA DEP, 1991.

Figure 11-1. (Continued).

Appendix A
Additional EPA Sources of Information on Composting

EPA Publications on Topics Relating to Composting

The following publications are available at no charge from the EPA RCRA/Superfund Hotline. Call 800-424-9346, or TDD 800-553-7672 for the hearing impaired, Monday through Friday, 8:30 a.m. to 7:30 p.m., EST. In Washington, DC, call 703-412-9810 or TDD 703-412-3323.

Decision-Maker's Guide to Solid Waste Management. EPA/530-SW-89-072. 1989.

Markets for compost. EPA/530-SW-90-073b. 1993.

Promoting Source Reduction and Recyclability in the Marketplace. EPA/530-SW-89-066. 1989.

Recycling Grass Clippings. EPA/530-F-92-012.

Residential Leaf Burning: An Unhealthy Solution to Leaf Disposal. EPA/452-F-92-007.

Sites for Our Solid Waste: A Guidebook for Effective Public Involvement. EPA/530-SW-90-019. 1990.

Yard Waste Composting: A Study of Eight Programs. EPA/530-SW-89-038. 1989.

Yard Waste Composting. EPA/530-SW-91-009.

The following publications are available from the National Technical Information Service (NTIS). Call 800-553-6847, Monday through Friday, 8:30 a.m. to 5:30 p.m. In Washington, DC, call 703-487-4650.

Characterization of Municipal Solid Waste in the United States. PB92-207 166. 1992.

Charging Households for Waste Collection and Disposal: The Effects of Weight- or Volume-Based Pricing on Solid Waste Management. PB91-111 484. 1990.

Variable Rates in Solid Waste: Handbook for Solid Waste Officials. PB90-272 063. 1990.

U.S. Environmental Protection Agency

Regional Offices

Region 1

U.S. EPA Region 1
J.F.K Federal Building
Boston, MA 02203
617-565-3420

Region 2

U.S. EPA Region 2
26 Federal Plaza
New York, NY 10278
212-264-2657

Region 3

U.S. EPA Region 3
841 Chestnut Building
Philadelphia, PA 19107
215-597-9800

Region 4

U.S. EPA Region 4
345 Courtland Street, NE
Atlanta, GA 30365
404-347-4727

Region 5

U.S. EPA Region 5
77 West Jackson Boulevard
Chicago, IL 60604-3507
312-353-2000

Region 6

U.S. EPA Region 6
First Interstate Bank Tower
1445 Ross Avenue
Dallas, TX 75202-2733
214-655-6444

Region 7

U.S. EPA Region 7
726 Minnesota Avenue
Kansas City, KS 66101
913-551-7000

Region 8

U.S. EPA Region 8
Denver Place (811WM-RI)
999 18th Street, Suite 500
Denver, CO 80202-2405
303-293-1603

Region 9

U.S. EPA Region 9
75 Hawthorne Street
San Francisco, CA 94105
415-744-1305

Region 10

U.S. EPA Region 10
1200 Sixth Avenue
Seattle, WA 98101
206-553-4973

Appendix B
Composting Equipment

Different types of equipment are used during composting to collect and transport the feedstock materials, to remove noncompostable materials for recycling or disposal, to increase the rate at which materials compost, to improve the quality of the finished compost product, to improve worker safety and working conditions, and to prepare the finished compost for marketing. Although the same types of equipment can be used to compost both yard trimmings and MSW, in many cases certain types of equipment are more appropriate for one type of composting than the other.

This appendix discusses the wide variety of equipment that is available for use in composting operations. The types of equipment discussed are divided into the following categories:

- Yard trimmings feedstock collection equipment
- Debagging equipment
- Sorting/separation equipment
- Size reduction equipment
- Mixing equipment
- Turning equipment
- Process control equipment
- Odor control equipment

Yard Trimmings Feedstock Collection Equipment

A variety of equipment exists for the collection of yard trimmings for processing and disposal. In most communities, yard trimmings are collected at curbside or citizens transport their materials to a specified drop-off area or transfer station. The main types of equipment used are trash collection vehicles and storage containers. Because compactors and containers are so common, this equipment is not discussed here.

There are several types of equipment available for yard trimmings collection today: mechanical scoops, which use either a bucket-like system to scoop yard trimmings or pincer-like systems to grab yard trimmings; and vacuum machines, which suck leaves through a nozzle for collection (Barkdoll and Nordstedt, 1991). These types of equipment are briefly described below:

- *Bucket Attachments* - These are standard attachments that can be fitted to a front-end loader and are used to scoop up yard trimmings and place them into holding containers.

- *Pincer Attachments* - These attachments can be fitted to front-end loaders or skid/steer loaders. Pincer buckets grab, rather than scoop up, the yard trimmings and place them into holding containers, usually on dump trucks or garbage packers.

- *Self-Contained Mechanical Scoops* - These systems use a series of rotating paddles that scoop yard trimmings off the ground and onto a conveyor that carries the yard trimmings to dump trucks. Mechanical scoops are usually mounted on small tractor trucks.

- *Vacuum Loaders* - Vacuum pressure is used to suck leaves directly into a separate enclosed container, usually built onto dump trucks.

- *Vacuum Collectors* - These self-contained units include both the vacuum equipment and the collection/storage units.

For more information on yard trimmings feedstock collection equipment, see Table B-1.

Debagging Equipment

For yard trimmings and MSW placed in plastic bags for collection, some system must be used to release the feedstock materials from the plastic bags and to remove the plastic so that it does not interfere with the composting process or diminish the quality of the finished compost product. Although manual opening and removal of bags is acceptable and widely used, a wide variety of commercial debagging equipment is now available.

Table B-1. A comparison of yard trimmings collection equipment.

Type of Equipment	Cost	Major Advantages	Major Disadvantages
Bucket Attachments	Usually included in price of front-end loader.	Many public works agencies have them available; work well on hard surfaces.	Not very efficient for collecting loose yard trimmings; must be fitted to a front-end loader or similar vehicle; pick up dirt and gravel.
Pincer Attachments	$2,300 to $12,000.	Well suited for collecting yard trimmings, particularly leaves; good for wet leaves.	Must be fitted to a front-end loader or similar vehicle; might need street sweeper to follow, depending on type of pincer.
Self-Contained Mechanical Scoops	$85,000 to $100,000.	Well suited for collecting yard trimmings, particularly leaves; the unit is self-contained and no front-end loader is necessary.	Must be fitted to a front-end loader or similar vehicle.
Vacuum Loaders	$6,000 to $25,000.	Well suited for collecting leaves; can be detached from the collection vehicle to dump; can be mounted to the front of the collection/storage vehicle.	Must be mounted to a collection/storage vehicle; labor intensive.
Vacuum Collectors	$15,000 to $40,000.	Well suited for collecting leaves; the system is self-contained and includes a self-dumping collection unit along with the vacuum machine; a compactor is available through at least one manufacturer.	Not good for grass and leaves when they become wet or frozen; must be mounted to a collection/storage vehicle; labor intensive.

Source: Barkdoll and Nordstedt, 1991.

There are two general categories of commercially available debagging devices: slitter/trommel devices and augers. All of these debagging systems can be used for both yard trimmings and MSW. Debagging can occur at the facility or at curbside. All the equipment described below is employed at the facility, except for the compactor truck with auger, which is attached to a collection vehicle:

- *Slitter/Trommel Devices* - A wide variety of slitter and trommel equipment are commercially available today. With these systems, the bags are either fed directly into the slitter or are transported by conveyors to the slitter unit. Slitters generally use counter-rotating blades to slice open the bags. The bags and their contents then fall or are transported by conveyors into the trommel unit. Feedstock is screened from the bags in the trommel unit, either through vibrating action of flat screens or rotating action of drumlike screens. Bags are removed from the trommel by hand or by air or water classifying units. Slitter/trommel systems can be used in conjunction with separation devices to remove metals, plastics, glass, etc.

- *Augers* - With auger systems, bags are loaded into the auger unit where a sharp-edged, screw-like shaft rotates and slices open the bags. The bags are turned and mixed by the auger so that their contents are released. The auger units generally are on an angle with the infeed end higher than the discharge end. Gravity moves the bags and feedstock materials through these systems. The materials are released at the discharge end and bags are removed by hand or by classifiers.

- *Trash Compactor Trucks with Augers* - Although the primary purpose of these units is trash compaction, most bags loaded into these units break during processing. The bags are dropped into the unit and are ripped when they pass into the compactor. The turning of the auger further rips the bags, compacts the materials, and releases much of the material from the bags.

- *Spike and Conveyor Debagging Systems* - One company has developed a system where bags are loaded into a hopper where a spiked chain grabs and drags the bags into a trough with two counter-rotating wheels edged with vertical spikes. The bags press down on each other and the pressure causes the bags to be gripped by the spikes and ripped open by the counter-rotating motion. Contents of the bags spill onto a conveyor and the bag clings to the spikes. A vacuum machine removes the bags from the spikes.

- *Specially Designed Windrow Turners* - The elevating face of these windrow turners lifts the plastic bags with paddle-like extensions. The bags are hooked by trencher teeth on the front of the windrow turner and ripped open. As the bags flip over the top, their contents are spilled out and the bags remain hung on the teeth. Common systems can be adapted with a bar containing cutting blades to enhance bag opening and with spiked teeth (rather than the normal cup-like teeth) to increase bag retrieval. Certain windrow turners also can be adapted to keep material in the windrow away from the bearings and can be fitted with a radial arm to cut bags off the drum of the windrow turner when they get wrapped around it.

- *Mechanical Jaw Debagging Systems* - Front-end loaders or in-line conveyors are used to feed bags into these systems. When the upper jaws of the unit open, the bags fall into the processing area. When the upper jaws close, new bags cannot enter the processing unit until processing of the original bags is completed. Bags are held in a fixed position while rippers slash them open. The lower jaws open and drop the material and bags onto a conveyor. A system is being developed to mechanically remove the bags, but currently manual separation of the bags is required.

- *Saw-Toothed Blade Debagging Systems* - These relatively small units can be used as stationary systems or they can be pulled by tractors. Power is supplied from the tractors, or the units can be adapted for electric motors. Bags are manually fed onto a conveyor, which is at a 45° angle. The conveyor is equipped with heavy, metal bars that are perpendicular to the conveyor and spaced 18 inches apart. Each bar has two tines that hook the bags. Hooked bags must pass under a saw-toothed blade, which tears them open. At the top of the conveyor, the contents of the bag are dropped into a bin. A blower blows the materials from the bags into a hopper or a truck or directly into a windrow for mobile systems. The bags stay attached to the tines and dangle down until they are caught by a double roller that pulls them from the tines and feeds them into a baler.

For more information on debagging equipment, see Table B-2.

Sorting/Separation Equipment

Sorting and separation of both yard trimmings and MSW usually are warranted to remove noncompostable materials and contaminants from the compost feedstock. A variety of sorting systems are available, ranging from technologically simple and labor intensive methods like manual removal of noncompostables and contaminants from a conveyor to technologically complex systems that mechanically separate noncompostables from compostables on the basis of physical characteristics such a weight, size, conductivity, and magnetic properties. Although all sorting/separation equipment can be used for both yard trimmings and MSW feedstock, certain types of equipment are more appropriate for one type of composting than another. The main types of sorting/separation equipment are briefly defined below:

- *Conveyors* - Conveyors are mechanical systems with belts that slowly pass over rotating wheels. Conveyor belts are used in the sorting/separation phase of composting to allow a constant stream of feedstock to pass by workers who manually remove noncompostables and other contaminants. The conveyor belt must be narrow enough for the workers to reach its center. Conveyors are needed primarily at MSW composting facilities.

- *Screens* - There are many types of screens, but all sort materials based on their size. The following types of screens are used in yard trimmings and MSW composting (Richard, 1992; Rynk et al., 1992):

  - *Stationary screens* - These are grates that are held in place while feedstock materials are dropped onto them. They retain materials that are larger than the mesh on the grate, while materials that are smaller than the mesh fall through. Screens with different mesh sizes can be positioned to separate materials into different size categories.

  - *Shaker screens* - Mechanical action causes these screens to move with an up and down motion. This movement helps to sift the materials through the mesh on the screens. The motion minimizes blinding. Heavy balls can be placed on the screen to help dislodge materials that are clogging the screen. Screens with different mesh sizes can be used with shaker screens to separate materials into different sizes.

  - *Vibrating screens* - These are similar to shaker screens except that the rate of motion is much more rapid. Vibrating screens are placed on an angle to remove oversized materials. Like shaker screens, different mesh sizes and cleaning balls can be used.

  - *Trommel screens* - These are long, cylindrical screens that are placed on an angle so that materials flow through them. Materials that are smaller than the grate fall through. As trommel screens rotate, a brush is passed over the top of the screen to remove lodged materials and prevent clogging of the screen. Trommel screens can separate items of different sizes by having a mesh gradient that increases away from the infeed end of the screen.

Table B-2. A comparison of debagging equipment.

Type of Equipment	Efficiency	Cost	Major Advantages	Major Disadvantages
Slitter/Trommel Devices	95% of the bags are opened; 75 to 99% of the bag contents are removed; 15 to 40 tons of material are processed per hour (some systems can process up to 90 tons per hour for just yard trimmings); 1,700 bags per hour can be processed.	$90,000 to $270,000.	Bags are left whole or in large pieces; a wide variety of systems are commercially available with different adaptations for specific requirements.	Manual separation or another bag removal mechanism must be used; up to 30% of the shredded plastic or paper bag pieces can remain in composting material, making a screening step necessary.
Augers	Approximately 98% of bag contents are removed; up to 25 tons of material can be processed per hour.	$65,000 to $75,000.	Models are available with augers that reverse direction when jammed; bags are left whole or in large pieces; companion baling systems for bags and separation devices are available.	Small bags can squeeze through the system without being opened; manual separation of bags is required.
Trash Compactor Trucks with Augers	Data not available.	Approximately $69,000.	Excellent safety features, including automated lifting of carts and the auger compaction combine, which reduce injuries to collection personnel.	To be efficient enough for use with a composting operation, only paper bags can be processed; with plastic bags, further screening is required to remove bags.
Spike and Conveyor Debagging Systems	2,000 bags are opened per hour; approximately 10 tons of material can be processed per hour.	Approximately $95,000.	After processing, bags are whole or in large pieces; a vacuum component removes the bags and there is no need for manual separation; virtually all bags are retrieved with this method.	It will be necessary to customize the system to tailor it to a specific facility.
Specially Designed Windrow Turners	Approximately 90% of the bags were removed with three passes of the windrow turner with one model investigated; with another model, 80% of the bags were removed with four passes; approximately 41 tons of material can be processed per hour; 1,172 bags were opened per hour with one of the systems.	Approximately $57,000; $10,000 to $15,000 to retrofit certain models with a radial arm to remove bags wrapped around the drum.	The unit also can be used for windrow turning; some windrow turners can be purchased or retrofitted with a radial arm for removing bags that become wrapped around the windrow turner drum.	For maximum debagging efficiency, bags should only be placed three deep in a windrow; plastic bags can become wrapped around the drum of the windrow turner.
Debagging Attachments for Compactor Trucks	100% of the bags are removed.	Approximately $8,750.	All bags are removed; less labor and handling are required at the composting site, because all bags have been removed; can be mounted on any rear-load compactor truck.	Labor is required to hold the bags in place while they are being processed; a rear-load compactor truck is needed to use this system; it is only appropriate for small communities with 10,000 to 25,000 residents.
Mechanical Jaw Debagging Systems	90% of the bags are opened; 99% of the bag contents are removed; 1,200 to 1,500 bags per hour can be processed.	Approximately $49,500.	Removes almost all bag contents.	Must be fed by conveyor or front-end loader.
Saw-Toothed Blade Debagging Systems	1,200 bags per hour can be processed.	Approximately $88,000.	Can be used as a stationary unit or pulled by a tractor; self adjusts for different sized bags; with mobile units, material can be blown from the bags directly into the windrows; a mechanical device automatically bales the bags.	Must be fed by conveyor.

Source: Ballister-Howells, 1992.

- *Disc screens* - These systems consist of many rotating scalloped-shaped, vertical discs. Small items fall through the spaces between the discs, and large items are moved over the discs to the discharge end of the system. These systems remove large items but do not separate the smaller pieces by size.

- *Rotary screens* - Feedstock is loaded onto spinning, perforated discs with this system. Oversized materials are thrown from the screen because of the spinning action. Undersized materials fall through the perforations in the discs.

- *Flexing belt screens* - Belts with slots or some other type of perforation are used with these systems. Segments of the belt are flexed and snapped in an alternating pattern, or the belt moves with a wave-like motion. This movement helps undersized materials to fall through the belt and removes materials that are clogging the screen.

- *Auger and trough screens* - These systems use a perforated trough to screen materials. An auger rotates in the trough, helping fine material fall through the perforations and moving oversized material out of the trough. Auger and trough screens with perforations of different sizes can be used to separate materials by size. This type of screen is primarily used to sort fine materials from wood chips.

- *Magnetic Recovery Systems* - With these systems, a magnetic field removes ferrous metals from the rest of the feedstock material. The following types of magnetic separators are commonly used with yard trimmings and MSW composting systems:

  - *Overhead belt magnets* - Cylindrical magnets are installed over a conveyor belt, which carries feedstock. A belt is secured around the magnets, which rotate to move the belt. The belt is made of a material that becomes magnetized by the magnets, allowing the belt to attract ferrous metals and remove them from the conveyor belt below. The magnetized belt is either positioned directly over the conveyor belt or perpendicular to the conveyor belt. Generally, the magnetized belt moves more quickly than the conveyor belt to improve the efficiency of the magnetic separation.

  - *Drum magnets* - Drum magnets are placed over a conveyor at the end of a mechanism used to feed the separation system. Ferrous metals in the feedstock that pass under the rotating drum are attracted to the magnet and stick to the drum. An operation must be conducted to periodically scrape the ferrous metals from the drum.

- *Eddy-Current Separation Systems* - These systems are used to separate nonferrous metals from feedstock materials. A high-energy electromagnetic field is created, which induces an electrical charge in materials that conduct electricity, primarily nonferrous metals. The charge causes these materials to be repelled from the rest of the feedstock materials.

- *Air Classifiers* - With this technology, feedstock materials are fed through an air column at a specified rate. The air column is created by a vacuum that sucks light materials into a cyclone separator. As materials lose velocity in the cyclone, they are separated out by volume. Heavy materials are not even picked up by the sucking action and fall directly though. Air classifiers target light objects like paper and plastic and heavy objects like metals, glass, and organics.

- *Wet Separation Systems* - These systems use water rather than air to separate materials. Materials enter a circulating water stream. Heavy materials drop into a sloped tank, some of which vibrate. The heavy items then fall into an area where they can be removed. The lighter materials float and are removed from the water with stationary or rotating screens. These systems target organics and other floatable materials and sinkable materials like metal, glass, gravel, etc.

- *Ballistic or Inertial Separation Systems* - These separators are based on the density and elasticity characteristics of the feedstock materials. They use rotating drums or spinning cones to generate a trajectory difference that bounces heavy materials away from lighter materials. These systems separate materials into three categories: light materials, such as plastic and undecomposed paper; medium materials, such as compost; and heavy materials, such as metals, glass, gravel, etc.

For more information on sorting/separation equipment, see Table B-3.

Size Reduction Equipment

Size reduction of feedstock materials is done with both yard trimmings and MSW composting, primarily to increase the surface area to volume ratio of the material to speed up the composting process. Size reduction also can improve the effectiveness of certain sorting/separation technologies. Although the available size reduction equipment can be used for both yard trimmings and MSW

Table B-3. A comparison of sorting/separation equipment.

Type of Equipment	Major Advantages	Major Disadvantages
Conveyors	Relatively low cost; enables separation of all categories of materials.	Requires manual separation of materials.
Stationary Screens	Lack of mechanization makes them relatively inexpensive; screens of different mesh sizes can be used to sort materials into different size categories.	Screens easily become blinded; only separates by size and does not remove small pieces of glass, metal, plastic, and other noncompostables.
Shaker Screens	Screens of different mesh sizes can be used to sort materials into different size categories; movement and use of cleaning balls limits clogging of the screens.	Only separates by size and does not remove small pieces of glass, metal, plastic, and other noncompostables; mechanization increases expense.
Vibrating Screens	Some models have been adapted specifically for compost use; screens of different mesh sizes can be used to sort materials into different size categories; slope of screen helps move oversized materials to discharge point; movement and use of cleaning balls limits clogging of the screens.	Only separates by size and does not remove small pieces of glass, metal, plastic, and other noncompostables; mechanization increases expense.
Trommel Screens	A screen of varying mesh size can be used to sort materials into different size categories; slope of unit helps move oversized materials to discharge point; movement and use of cleaning brush limits clogging of the screen.	Only separates by size and does not remove small pieces of glass, metal, plastic, and other noncompostables; mechanization increases expense.
Disc Screens	Targets and eliminates large items; long history of use in other industries.	Only targets and eliminates large items; does not sort materials by size; does not remove small pieces of glass, metal, plastic, and other noncompostables; mechanization increases expense.
Rotary Screens	Movement helps limit clogging of the screen.	Only targets and eliminates large items; does not sort materials by size; does not remove small pieces of glass, metal, plastic, and other noncompostables; mechanization increases expense.
Flexing Belt Screens	Movement, particularly snapping and wave action, helps limit clogging of the screen.	Only targets and eliminates large items; does not sort materials by size; does not remove small pieces of glass, metal, plastic, and other noncompostables; mechanization increases expense.
Auger and Trough Screens	Troughs with perforations of varying sizes can be used to sort materials into different size categories; movement of the auger helps move oversized materials to discharge point; movement limits clogging of the screen; designed to remove wood chips from finer materials.	Only separates by size and does not remove small pieces of glass, metal, plastic, and other noncompostables; mechanization increases expense.
Overhead Belt Magnets	Very effective at separating ferrous metals from the rest of the feedstock materials; relatively inexpensive system for separating ferrous metals.	Can only be used to separate ferrous metals from the rest of the feedstock materials; relatively ineffective for feedstock placed on conveyors in thick layers; a second belt is required.
Drum Magnets	Very effective at separating ferrous metals; relatively inexpensive; a second belt is not required.	Relatively ineffective for feedstock placed on conveyors in thick layers.
Eddy Current Separation Systems	Effective at recovering nonferrous materials (these cannot be separated or recovered with traditional magnet systems).	If magnetic separation is not conducted prior to this process, high levels of contamination with ferrous metals occurs; can only be used to separate nonferrous metals from the rest of the feedstock materials; relatively ineffective for feedstock placed on conveyors in thick layers.
Air Classifiers	Light materials that are larger in size (such as plastic and paper) can be removed.	Only targets and eliminates relatively light or heavy items; does not remove medium-weight noncompostables; mechanization increases expense.

Table B-3. (Continued).

Type of Equipment	Major Advantages	Major Disadvantages
Wet Separation Systems	Particularly effective at removing organics because they float; allows heavy, sharp objects (such as glass pieces) to be safely removed.	Size reduction is needed before this technology is used; only targets and separates relatively light materials; does not remove lightweight noncompostables.
Ballistic or Inertial Separation Systems	Sorts and separates inorganics (glass, metal, and stone fall into separate bins); can use lasers or optical scanners to target certain inorganics, improving recovery rate.	Only targets and eliminates relatively dense items; does not remove dense noncompostables; mechanization increases expense.

Source: Rynk et al., 1992; Richard, 1992; Glaub et al., 1989.

composting, certain types of equipment are preferable depending on the type of feedstock. In the following list, the most common types of size reduction equipment available for use with yard trimmings and MSW composting are briefly described:

- *Hammermills* - With these systems, either free-swinging hammers strike and crush the feedstock materials or the feedstock materials are ground against fixed hammers and broken into smaller pieces. Hammermills must be well ventilated to prevent explosions that could arise from clogging. The following types of hammermills are most commonly used for composting:

  - *Horizontal hammermills* - These systems use counter-rotating hammers to crush feedstock materials. The free-swinging hammers are attached to horizontal shafts. Size-reduced feedstock must pass through a grate before exiting the system.
  - *Vertical hammermills* - These systems are similar to horizontal hammermills, except that the free-swinging hammers are attached to vertical shafts.
  - *Flail mills* - With these hammermills, size reduced materials do not have to pass through a grate before exiting the system.
  - *Tub grinders* - This type of size reduction equipment is used primarily for yard trimmings. Feedstock materials are loaded into the tub, which rotates and moves the material across a fixed floor that holds the hammers. The movement of the tub grinds feedstock against the hammers.

- *Shear Shredders* - These systems use either fixed or free-swinging knives to slice feedstock materials into smaller sizes. Shredders typically require little maintenance.

  - *Fixed-knife shear shredders* - With these shredders, a cleated belt is used to force feedstock materials against fixed knives. The materials are raked and shredded by the movement. With this type of equipment, adjustable fingers catch oversized materials and push them back into the shredder. Glass items are rejected and fall through a trash chute.
  - *Rotating-knife shear shredders* - This type of shredder has two shafts with hooked cutter discs attached to them. The shafts are counter rotating and the discs interconnect. The discs slice the materials until they are small enough to fall through the spaces between the discs. The size of the reduced materials is dependent on the size of the cutter discs.

- *Rotating Drums* - These systems consist of a rotating cylinder that is positioned at an angle. Materials are fed into the drum and the rotating motion causes them to tumble around the cylinder. The tumbling action breaks up the materials as dense and abrasive items pulp the softer materials.

For more information on size reduction equipment, see Table B-4.

Mixing Equipment

Mixing is performed in both yard trimmings and MSW composting operations to optimize several characteristics of the composting feedstock such as moisture content, carbon-to-nitrogen ratios, pH, and particle size. Mixing can be done when the compost piles or windrows are being turned, which does not necessarily require special mixing equipment (this probably depends on feedstock and odor concerns, however). When more complete mixing is warranted, special mixing equipment can be obtained. This mixing equipment can be used for composting both yard trimmings and MSW. Because of the expense involved, however, mixing equipment tends to be used more frequently for MSW composting because the heterogeneity of these feedstock increases the need for mixing before composting.

Table B-4. A comparison of size reduction equipment.

Type of Equipment	Capacity	Cost	Major Advantages	Major Disadvantages
Hammermills	4 to 75 tons per hour (or 60 to 450 cubic yards per hour, depending on the measure used).	$14,000 to $450,000.	Tend to reduce materials into smaller sizes than other size-reduction equipment.	Care must be taken in selecting an appropriate hammermill for MSW; create more noise than other types of size reduction equipment.
Tub Grinders	5 to 50 tons per hour (or 80 to 100 cubic yards per hour, depending on the measure used).	$20,000 to $191,400.	A wide variety of tub grinders are available; portable or stationary units are available.	Can require careful maintenance.
Shear Shredders	0.4 to 110 tons per hour (or 50 to 250 cubic yards per hour, depending on the measure used).	$11,000 to $360,000.	A wide variety of shear shredders are available; materials tend to be torn apart, which opens up their internal structure and speeds the composting process; often can be mounted on a trailer.	Thin, flexible items (like plastic sheeting) might not be cut or torn; might not be able to process oversized equipment.
Rotating Drums	One model claims 75 tons per hour.	$135,000.	Materials are mixed while being size-reduced.	Actual size reduction varies with feedstock mix; long noncompostable items (like plastic sheeting and cables) usually must be manually removed.

Source: Barkdoll and Nordstedt, 1991.

Some facilities use the same equipment for size reduction and mixing. Mixing equipment is typically divided into batch systems and flow-through systems. Batch systems work with one load of material at a time. They are usually mounted on a truck or wagon so that mixed material can be placed directly on the windrow or composting pile. Flow-through systems are always stationary. Usually fed and emptied with a conveyor, they can process a continuous stream of material. Both types of mixing systems blend material by employing one of the technologies (or a combination of the technologies) described below:

- *Auger Mixers* - These consist of one or a number of rotating screws that chop, turn, and mix materials; used primarily in batch systems.

- *Barrel Mixers* - These mixers use paddles attached to a rotating shaft to stir material. Material is continuously fed into a vertical or inclined stationary drum; used primarily in flow-through systems.

- *Drum Mixers* - These are slowly turning, inclined drums that tumble and blend material. Sometimes the drums are divided into chambers for each stage of the mixing process.

- *Pugmill Mixers* - These mixers blend material with hammers attached to counter-rotating shafts; used primarily in flow-through systems.

For more information on mixing equipment, see Table B-5.

Turning Equipment

Because large quantities of feedstock materials must be handled, even with small composting operations, some type of equipment is needed to turn compost piles or windrows with almost any municipal composting operation. This equipment can range from machinery not specifically meant for composting operations, such as front-end loaders, to highly specific types of windrow turners. The same types of equipment can be used to compost both yard trimmings and MSW. The following is a list of the most common types of turning equipment used in composting operations:

- *Front-End Loaders* - These vehicles have a shovel-like attachment at the front of the machine. The attachment can be raised by a hydraulic mechanism to lift feedstock materials and tipped to release the materials into piles or windrows.

- *Bucket Loaders* - These loaders are similar to front-end loaders except that the attachment used to raise and tip the feedstock materials is bucket-shaped.

- *Manure Spreaders* - With these vehicles, feedstock is loaded in a hopper at the rear of the cab. Rotating paddles push materials out of the back of the storage

Table B-5. A comparison of mixing equipment.

Type of Equipment	Major Advantages	Major Disadvantages
Batch Mixers	After mixing, the materials can be discharged directly into a composting pile or windrow; most mixers can be mounted on an available truck or wagon; good for smaller facilities.	If the mixer is operated for too long, compaction occurs; fibrous materials, such as straw, can wrap around the mixing mechanism; limited capacity.
Barrel Mixers	High capacity because of continuous operation; good for large facilities.	Do not significantly compress materials; high capital costs.
Pugmill Mixers	Achieves best size reduction; produces high-quality mix.	Maintaining hammers can be costly.
Drum Mixers	Facilitates composting since microbial decomposition can begin in drum.	Wet material might stick to drum at high speeds or form clumps at low speeds.
Auger Mixers	When used in batch system, materials can be moved to curing pad or windrow while being mixed; produces uniform mix.	Can shred materials and therefore reduce the effectiveness of bulking agents.

Source: Higgins et al., 1981.

container, mixing the materials as they are released. The materials can be released while the spreader is in a stationary position into a pile or while the spreader is slowly moving.

- *Tractor/Trailer-Mounted Windrow Turners* - These turners must be pushed or pulled by a tractor or another vehicle. They ride on the side of the vehicle and rotating paddles or other extensions flip and turn the material in the windrow.

- *Tractor-Assisted Windrow Turners* - These turners are similar to tractor/trailer-mounted windrow turners except that they require the tractor to provide a power source to rotate their turning mechanism. The tractor must have a power gear or hydrostatic drive to power the turners.

- *Self-Driven Windrow Turners* - A wide variety of self-driven, self-powered turners exist. Some models have turning mechanisms that ride to the side of the vehicle. Others straddle the windrow while the turning mechanism flips and turns the composting materials.

For more information on turning equipment see Table B-6.

Process Control Equipment

Two of the factors most commonly controlled with compost operations are temperature and oxygen levels. Turning of windrows and compost piles is a common way to control these factors. Specially designed forced aeration equipment is available to control temperature and oxygen levels in compost piles and windrows. The primary categories of forced aeration equipment are as follows:

- *Suction Systems* - A vacuum device is used to draw air through the composting mass. The air is collected in an exhaust pipe and can be treated for odor control. Leachate also is removed.

- *Positive Pressure Systems* - With this equipment, a blower pushes air into the composting mass.

Three types of methods can be used to control the aeration of the composting mass. These are described below:

- *Continuous Aeration* - With these systems, aeration devices are run without interruption (although they can be turned off manually).

- *Timer Control* - With these systems, empirical data is gathered to determine when and for how long forced aeration equipment should be run. Timers are then used to turn the aeration equipment on and off.

- *Automatic Feedback Control* - With these systems, temperature or oxygen monitoring equipment is used to determine when critical levels of these parameters have been reached. When a critical level has been reached, the sensors trigger a mechanism that turns the aeration equipment on or off.

For more information on process control equipment, see Table B-7.

Odor Control Equipment

Numerous odor control methods are used at composting facilities, ranging from simple and inexpensive procedures (such as adding wood ash to the compost or increasing dilution of compost exhaust air with ambient air by installing fans or raising stack height) to the more complex and costly equipment discussed below. Appropriate odor control methods will vary for different facilities depending on the type and amount of control needed and on financial resources.

- *Biofilters* - The exhaust air from the compost process is passed through a biological filter medium, such as soil or sand. The air is evenly distributed through the medium by either an open system, consisting of perforated pipes set in gravel over which the biofilter medium is placed, or by a closed system, consisting of a vessel (with a perforated aeration plenum) filled with the biofilter medium. Odorous compounds in the exhaust air are removed by the biofilter through various physical, biological, and chemical processes. For example, odorous compounds are broken down into non-odorous materials such as carbon dioxide, water, and nitrogen, or are absorbed or adsorbed by the biofilter. Some researchers have recommended

Table B-6. A comparison of turning equipment.

Type of Equipment	Capacity	Cost	Major Advantages	Major Disadvantages
Loaders	$3 \frac{1}{4}$ to 4 yd^3 bucket.	$120,000 to $170,000.	Readily available in many municipalities; self-powered and self-driven; materials are not loaded into the vehicle; can be fitted with buckets or other attachments according to facility needs.	A space the width of the loader is required between every pair of windrows; poor mixer.
Manure Spreaders	300 to 350 bushel loads.	$9,500 to $11,000.	Mixes materials thoroughly.	Materials must be loaded into the spreader before they can be turned; a space the width of the spreader (and the vehicle that is used to load materials into the spreader) is required between every pair of windrows; it takes significantly more time to conduct the mixing operation than with other equipment alternatives.
Tractor/Trailer-Mounted Windrow Turners	300 to 3,000 tons per hour.	$15,000 to $100,000.	Very efficient for turning windrows and mixing materials; self-powered; a variety of models exist with different turning mechanisms.	Must be mounted to a tractor or another vehicle; for most models, a space the width of the tractor and the turner is required between each windrow or pile; for single-pass turner models, however, a space the width of the tractor and the turner is required between every pair of windrows.
Tractor-Assisted Windrow Turners	300 to 1,200 tons per hour.	$7,400 to $68,000.	Very efficient for turning windrows and mixing materials; a variety of models exist with different turning mechanisms.	Requires a separate power source; must be mounted to a tractor or another vehicle; for most models, a space the width of the tractor and the turner is required between each windrow or pile; for single-pass turner models, however, a space the width of the tractor and the turner is required between every pair of windrows.
Self-Driven Windrow Turners	1,000 to 4,000 tons per hour.	$89,000 to $250,000.	Very efficient for turning windrows and mixing materials; self-driven; self-powered; a variety of models exist with different turning mechanisms; some models straddle the windrows and require minimal space between windrows.	For some models, a space the size of the turner is required between every windrow or pair of windrows.

Source: Barkdoll and Nordstedt, 1991.

Table B-7. A comparison of process control equipment.

Type of Equipment	Capacity	Major Advantages	Major Disadvantages
Suction Systems	Medium.	Exhaust gas can be captured and treated to control odors.	Water vapor must be removed from the exhaust gas before it reaches the suction device; continuous use can lead to variable temperature, oxygen, and moisture levels in the compost pile or windrow.
Positive Pressure Systems	Medium.	Provides more efficient and uniform aeration than suction devices.	Odor prevention is difficult; continuous use can lead to variable temperature, oxygen, and moisture levels in the compost pile or windrow; tends to create an unpleasant working environment.
Continuous Aeration	Low.	Lower airflow rates are required; no timer or feedback mechanism is required.	Variable temperature and oxygen levels that are disruptive to the composting process are likely inside the composting pile or windrow.
Timer Control	Medium.	More uniform temperature or oxygen levels can be achieved than with continuous aeration systems.	Temperatures are not necessarily maintained at optimal levels; experimentation is needed to determine the best time schedule for aeration.
Automatic Feedback Control	High.	Relatively uniform temperature or oxygen levels in the compost pile or windrow can be achieved; optimal temperature or oxygen levels can be maintained.	More powerful aeration equipment is necessary.

Source: Richard, 1992; Rynk et al., 1992.

that further research, such as measurements of odor pervasiveness and intensity before and after air passes through the biofilter, be conducted to verify the odor removal efficiency of biofilters.

- *Wet Scrubbers* - Air from the composting process is exposed to a scrubbing solution, which reacts with and removes the odorous compounds in the air (e.g., through oxidation). Multistage scrubbers are generally needed to achieve adequate odor control. It is essential that chemical reactions in scrubbers occur in the correct sequence; otherwise the correct reactions may not occur, or other, odor-forming reactions might result. The two most common types of wet scrubbers are packed tower and mist scrubbers. Packed tower scrubbers pass the air through packing media through which the scrubbing solution circulates. Mist scrubbers atomize the scrubbing solution into droplets that are dispersed through the exhaust air stream.

- *Carbon Adsorption* - Air from the compost process enters a vessel containing beds of granular activated carbon and is dispersed across the face of the beds. The activated carbon adsorbs the odorous compounds in the air stream.

- *Thermal Regenerative Oxidation* - The compost air stream is exposed to temperatures of approximately 1,400 °F for one second. The high temperature reduces odors.

For more information on odor control equipment, see Table B-8.

Appendix B Resources

Barkdoll, A.W., and R.A. Nordstedt. 1991. Strategies for yard waste composting. BioCycle. May, 32(5):60-65.

Ballister-Howells, P. 1992. Getting it out of the bag. BioCycle. March, 33(3):50-54.

Glaub, J., L. Diaz, and G. Savage. 1989. Preparing MSW for composting. As cited in: The BioCycle Guide to Composting Municipal Wastes. Emmaus, PA: The JG Press, Inc.

Higgins, A.J., et al., 1981. Mixing systems for sludge composting. Biocycle. May, 22 (5):18-22.

Richard, T.L. 1992. Municipal solid waste composting: physical and biological processing. Biomass & Bioenergy. Tarrytown, NY: Pergamon Press. 3(3-4):195-211.

Rynk, R., et al., 1992. On-farm composting handbook. Ithaca, NY: Cooperative Extension, Northeast Regional Agricultural Engineering Service.

Table B-8. A comparison of odor control technologies.

Type of Equipment	Cost	Major Advantages	Major Disadvantages
Biofilters	Capital cost: $240,000 (1,600 CFM). Annual O&M costs: $140,000.	High removal rates at moderate cost.	Possible short-circuiting of exhaust gases; tendency of media to dry out, reducing effectiveness; need to maintain pH buffering capacity in media.
Multistage Wet Scrubbers			
Packed Tower	Capital cost: $1,000,000 (65,000 CFM). Annual O&M costs: $240,000 to $360,000.	Effective for ammonia removal; recirculation of solution enhances process efficiency.	Plugged media in packed towers; recirculation of solution may reintroduce odors into air stream.
Mist Scrubber	Capital cost: $1,000,000 (25,000 CFM). Annual O&M costs: $240,000 to $360,000.	"Once-through" passage of solution removes odorous compounds from air stream permanently.	Difficulty in maintaining effective chemical feedrates; plugged nozzles and filters.
Carbon Adsorption	Capital cost: $600,000. Annual O&M costs: $100,000.	Capable of removing a broad range of compounds.	Carbon capacity will be exhausted, requiring costly regeneration or replacement; thus standby unit is recommended; susceptible to plugging from particulates in air stream; recommended as secondary system only.
Thermal Regeneration	Capital cost: $1,500,000. Annual O&M costs: $240,000 to $360,000.	Recaptures heat, reducing fuel costs.	Numerous mechanical problems.

Source: Based primarily on estimates from pilot tests at the Concord, NH, biosolids composting facility (total exhaust air flow rate ranging from 24,000 CFM to 65,000 CFM), as reported in *Biocycle*, August 1992, and on personal communications with odor control researchers.

CFM = cubic feet/minute of air

Appendix C
Glossary of Compost Terms

actinomycetes - Family of microorganisms belonging to a group intermediary between bacteria and molds (fungi); a form of filamentous, branching bacteria.

aerated static pile - Composting system using controlled aeration from a series of perforated pipes running underneath each pile and connected to a pump that draws or blows air through the piles.

aeration (for composting) - Bringing about contact of air and composted solid organic matter by means of turning or ventilating to allow microbial aerobic metabolism (biooxidation).

aerobic - Composting environment characterized by bacteria active in the presence of oxygen (aerobes); generates more heat and is a faster process than anaerobic composting.

agricultural by-products or residuals - By-product materials produced from plants and animals, including manures, bedding, plant stalks, leaves, and vegetable matter.

air classification - The separation of materials using a moving stream of air; light materials are carried upward while heavy components drop out of the stream.

anaerobic - Composting environment characterized by bacteria active in the absence of oxygen (anaerobes).

bacteria - Unicellular or multicellular microscopic organisms.

bioaerosols - Biological aerosols that can pose potential health risks during the composting and handling of organic materials. Bioaerosols are suspensions of particles in the air consisting partially or wholly of microorganisms. The bioaerosols of concern during composting include actinomycetes, bacteria, viruses, molds, and fungi.

biochemical oxygen demand (BOD) - The amount of oxygen used in the biochemical oxidation of organic matter; an indication of compost maturity and a tool for studying the compost process.

biodegradability - The potential of an organic component for conversion into simpler structures by enzymatic activity.

biooxidation - Aerobic microbial metabolism of organic or inorganic compounds.

biosolids - Solid, wet residue of the wastewater purification process; a product of screening, sedimentation, filtering, pressing, bacterial digestion, chemical precipitation, and oxidation; primary biosolids are produced by sedimentation processes and secondary biosolids are the products of microbial digestion.

bulking agent - Material, usually carbonaceous such as sawdust or woodchips, added to a compost system to maintain airflow by preventing settlement and compaction of the compost.

carbon to nitrogen ratio (C:N Ratio) - Ratio representing the quantity of carbon (C) in relation to the quantity of nitrogen (N) in a soil or organic material; determines the composting potential of a material and serves to indicate product quality.

cation exchange capacity (CEC) - A routine measure of the binding potential of a soil; measures the soil's ability to remove negative ions from metals and other compounds, allowing the ions to form insoluble compounds and precipitate in the soil; determined by the amount of organic matter and the proportion of clay to sand—the higher the CEC, the greater the soil's ability to bind metals.

cellulose - Carbon component of plants, not easily digested by microorganisms.

co-composting - Composting process utilizing carbon-rich organic material (such as leaves, yard trimmings, or mixed municipal solid waste), in combination with a nitrogen-rich amendment such as biosolids.

compost - The stabilized product of composting which is beneficial to plant growth; it has undergone an initial, rapid stage of decomposition and is in the process of humification.

compostable - Organic material that can be biologically decomposed under aerobic conditions.

composting - The biodegradation, usually aerobic and thermophilic, that involves an organic substrate in the solid state; evolves by passing through a thermophilic stage with a temporary release of phytotoxins; results in the production of carbon dioxide, water, minerals, and stabilized organic matter.

composting, municipal - Management method whereby the organic component of municipal discards is biologically decomposed under controlled conditions; an aerobic process in which organic materials are ground or shredded and then decomposed to humus in windrow piles or in mechanical digesters, drums, or similar enclosures.

curbside pickup - The curbside collection and transport of used household materials to a centralized handling facility (municipal or private) such as a transfer station, a materials recovery facility (MRF), an incinerator, or landfill. Materials at curbside might be mixed together in common containers or source separated by the householder into separate fractions such as newspapers, glass, compostables, or any variation of mix and separation.

curbside recycling - Residents separate recyclables from their trash and leave the recyclables on their curbside for collection.

cured compost - A stabilized product that results from exposing compost to a prolonged period of humification and mineralization.

curing - Late stage of composting, after much of the readily metabolized material has been decomposed, which provides additional stabilization and allows further decomposition of cellulose and lignin.

decomposition - Conversion of organic matter as a result of microbial and/or enzymatic interactions; initial stage in the degradation of an organic substrate characterized by processes of destabilization of the preexisting structure.

denitrification - The biological reduction of nitrogen to ammonia, molecular nitrogen, or oxides of nitrogen, resulting in the loss of nitrogen into the atmosphere.

digester - An enclosed composting system with a device to mix and aerate the materials.

drop off - Individuals take recyclable materials to a recycling center.

drum composting system - Enclosed cylindrical vessel which slowly rotates for a set period of time to break up and decompose material.

endotoxins - A toxin produced within a microorganism and released upon destruction of the cell in which it is produced. Endotoxins can be carried by airborne dust particles at composting facilities.

enclosed system - See "in-vessel composting."

erosion - The removal of materials from the surface of the land by weathering and by running water, moving ice, and wind.

feedstock - Decomposable organic material used for the manufacture of compost.

finished product - Compost material that meets minimum requirements for public health, safety, and environmental protection and is suitable for use as defined by finished product standards.

food scraps - Residual food from residences, institutions, or commercial facilities; unused portions of fruit, animal, or vegetable material resulting from food production.

fungi - Saprophytic or parasitic multinucleate organisms with branching filaments called hyphae, forming a mass called a mycelium; fungi bring about cellulolysis and humification of the substrate during stabilization.

green materials - Portion of the municipal discards consisting of leaves, grass clippings, tree trimmings, and other vegetative matter.

hammermill - Machine using rotating or flailing hammers to grind material as it falls through the machine or rests on a stationary metal surface.

heavy metals - Elements having a high specific gravity regulated because of their potential for human, plant, or animal toxicity, including cadmium (Cd), copper (Cu), chromium (Cr), mercury (Hg), nickel (Ni), lead (Pb) and zinc (Zn).

household hazardous waste - Products containing hazardous substances that are used and disposed of by individuals rather than industrial consumers; includes some paints, solvents, and pesticides.

humus - A complex aggregate of amorphous substances, formed during the microbial decomposition or alteration of plant and animal residues and products synthesized by soil organisms; principal constituents are derivatives of lignins, proteins and cellulose; humus has a high capacity for cation exchange (CEC), for combining with inorganic soil constituents, and for water absorption; finished compost might be designated by the general term humus.

hydromulching - An application method using a water jet to spread a mulch emulsion on a land surface.

in-vessel composting - (also "enclosed" or "mechanical") A system using mechanized equipment to rapidly decompose organic materials in an enclosed area with controlled amounts of moisture and oxygen.

inerts - Nonbiodegradable products (glass, plastics, etc.).

inorganic - Substance in which carbon-to-carbon bonds are absent; mineral matter.

integrated waste management - The complementary use of a variety of practices to handle municipal solid waste safely and effectively; techniques include source reduction, recycling/composting, combustion, and landfilling.

land reclamation - The restoration of productivity to lands made barren through processes such as erosion, mining, or land clearing.

landfilling - The disposal of discarded materials at engineered facilities in a series of compacted layers on land and the frequent daily covering of the waste with soil; fill areas are carefully prepared to prevent nuisances or public health hazards, and clay and/or synthetic liners are used to prevent releases to ground water.

leachate - Liquid which has percolated through materials and extracted dissolved and suspended materials; liquid that drains from the compost mix.

macronutrients - Nutritive elements needed in large quantities to ensure normal plant development.

mature compost - Compost that has been cured to a stabilized state, characterized as rich in readily available forms of plant nutrients, poor in phytotoxic acids and phenols, and low in readily available carbon compounds.

mesophilic stage - A stage in the composting process characterized by bacteria that are active in a moderate temperature range of 20 to 45°C (68 to 113°F); it occurs later, after the thermophilic stage and is associated with a moderate decomposition rate.

metabolism - Sum of the chemical reactions within a cell or whole organism, including the energy-releasing breakdown of molecules (catabolism) and the synthesis of complex molecules and new protoplasm (anabolism).

micronutrients - Nutritive elements needed in small quantities for healthy plant development; trace elements.

microorganisms - Small living organisms only visible with a microscope.

moisture content - The mass of water lost per unit dry mass when the material is dried at 103°C (217°F) for 8 hours or more. The minimum moisture content required for biological activity is 12 to 15 percent; it generally becomes a limiting factor below 45 to 50 percent; expressed as a percentage, moisture content is water weight/wet weight.

mulch - Any suitable protective layer of organic or inorganic material applied or left on or near the soil surface as a temporary aid in stabilizing the surface and improving soil microclimactic conditions for establishing vegetation; mulch reduces erosion and water loss from the soil and controls weeds.

municipal solid waste (MSW) - Discarded material from which decomposable organic material is recovered for feedstock to make compost. Municipal solid waste originates from residential, commercial, and institutional sources within a community.

nematodes - Elongated, cylindrical, unsegmented worms; includes a number of plant parasites (a cause of root damage) and human parasites.

nitrification - The oxidation of ammonia to nitrite and nitrite to nitrate by microorganisms.

organic - Substance that includes carbon-to-carbon bonds.

organic contaminants - Synthetic trace organics include pesticides and polychlorinated biphenyls (PCBs).

organic matter - Portion of the soil that includes microflora and microfauna (living and dead) and residual decomposition products of plant and animal tissue; any carbon assembly (exclusive of carbonates), large or small, dead or alive, inside soil space; consists primarily of humus.

organic soil conditioner - Stabilized organic matter marketed for conditioning soil structure; it also improves certain chemical and biological properties of the soil.

oxidation - Energy-releasing process involving removal of electrons from a substance; in biological systems, generally by the removal of hydrogen (or sometimes by the addition of oxygen); chemical and/or biochemical process combining carbon and oxygen and forming carbon dioxide (CO_2).

pathogen - An organism, chiefly a microorganism, including viruses, bacteria, fungi, and all forms of animal parasites and protozoa, capable of producing an infection or disease in a susceptible host.

persistence - Refers to a slowly decomposing substance which remains active in the natural cycle for a long period of time.

pH - The negative logarithm of the hydrogen ion concentration of a solution, a value indicating the degree of acidity or alkalinity; pH 7 = neutral, pH <7 = acid, pH >7 = alkaline (basic).

phytotoxic - Detrimental to plant growth; caused by the presence of a contaminant or by a nutrient deficiency.

polychlorinated biphenyls (PCBs) - A class of chlorinated aromatic hydrocarbons representing a mixture of specific biphenyl hydrocarbons which are thermally and chemically very stable; some PCBs are proven carcinogens.

putrescible waste - Organic materials prone to degrade rapidly, giving rise to obnoxious odors.

recyclable - Products or materials that can be collected, separated, and processed to be used as raw materials in the manufacture of new products.

recycling - Separating, collecting, processing, marketing, and ultimately using a material that would have been thrown away.

runoff - Water that flows over the earth's surface that is not absorbed by soil.

screening - The sifting of compost through a screen to remove large particles and to improve the consistency and quality of the end product.

shredder - Mechanical device used to break materials into small pieces.

size reduction - Generic term for separation of the aggregate or for breaking up materials into smaller pieces through abrasion, thermal dissociation, tearing, screening, tumbling, rolling, crushing, chipping, shredding, grinding, shearing, etc.; the process makes materials easier to separate and can increase surface area for composting.

soil amendment/soil conditioner - Soil additive which stabilizes the soil, improves resistance to erosion, increases permeability to air and water, improves texture and resistance of the surface to crusting, eases cultivation, or otherwise improves soil quality.

source reduction - The design, manufacture, purchase, or use of materials to reduce their amount or toxicity; because it is intended to reduce pollution and conserve resources, source reduction should not increase the net amount or toxicity generated throughout the life of the product; techniques include reusing items, minimizing the use of products that contain hazardous compounds, using only what is needed, extending the useful life of a product, and reducing unneeded packaging.

source separation - Separating materials (such as paper, metal, and glass) by type at the point of discard so that they can be recycled.

stability - State or condition in which the composted material can be stored without giving rise to nuisances or can be applied to the soil without causing problems there; the desired degree of stability for finished compost is one in which the readily decomposed compounds are broken down and only the decomposition of the more resistant biologically decomposable compounds remains to be accomplished.

stabilization - Stage in composting following active decomposition; characterized by slow metabolic processes, lower heat production, and the formation of humus.

static pile system - An aerated static pile with or without a controlled air source.

thermophilic stage - A stage in the composting process characterized by active bacteria that favor a high temperature range of 45 to 75°C (113 to 167°F); it occurs early, before the mesophilic stage, and is associated with a high rate of decomposition.

tilth - The physical state of the soil that determines its suitability for plant growth taking into account texture, structure, consistency, and pore space; a subjective estimation, judged by experience.

topsoil - Soil, consisting of various mixtures of sand, silt, clay, and organic matter, considered to be the nutrient-rich top layer of soil that supports plant growth.

toxicity - Adverse biological effect due to toxins and other compounds.

vector - Animal or insect—including rats, mice, mosquitoes, etc.—that transmits a disease-producing organism.

volatilization - Gaseous loss of a substance to the atmosphere.

windrow system - Elongated piles or windrows aerated by mechanically turning the piles with a machine such as a front-end loader or specially designed equipment.

wood scrap - Finished lumber, wood products and prunings, or stumps 6 inches or greater in diameter.

yard trimmings - Grass clippings, leaves, brush, weeds, Christmas trees, and hedge and tree prunings from residences or businesses.

Appendix C Resources

Composting Council. 1991. Compost facility planning guide. Washington, DC: Composting Council.

For Product Safety Concerns and Information please contact our EU representative GPSR@taylorandfrancis.com Taylor & Francis Verlag GmbH, Kaufingerstraße 24, 80331 München, Germany